W9-CEL-895

CHANCE

CHANCE

A Guide to Gambling, Love, the Stock Market, & Just About Everything Else

Amir D. Aczel

THUNDER'S MOUTH PRESS
NEW YORK

CHANCE

A GUIDE TO GAMBLING, LOVE,
THE STOCK MARKET, & JUST ABOUT EVERYTHING ELSE

Published by
Thunder's Mouth Press
An Imprint of Avalon Publishing Group Inc.
245 West 17th St., 11th Floor
New York, NY 10011

Library of Congress Cataloging-in-Publication Data is available.

ISBN 1-56858-316-8

9 8 7 6 5 4 3 2 1

Book design by Ink, Inc., New York.
Printed in the United States of America
Distributed by Publishers Group West

CONTENTS

INTRODUCTION

We have built a thousand temples to Fortune
and not one to Reason.

—Marcus Cornelius Fronto,
tutor to Marcus Aurelius

THE TWIN FORCES OF CHANCE AND MISCHANCE have beguiled humanity like none other. Why does fortune smile on some people, and smirk on others? What is luck, and why does it so often visit the undeserving? How can we predict the random events happening around us? Even better—how can we manipulate them?

It's easy to think of chance as beyond us. Early gods, astrology, superstition: all were attempts to explain the unexplainable. *Our village was devastated by drought because the goddess of luck turned her back on us. Her child grew up mad because he was born under a portentous sign. His slaves revolted and sacked his villa because he spilt salt at the banquet last night.* Today we might question these theories—but at least we must credit their creators with trying; they looked around themselves and refused to believe their world could be arbitrary. On a very fundamental level, chance may remain a mystery, but using the language of math we can now rename it "probability;" and now we can define it, and devise equations that account for it. Like sailors at sea, who may not be able to control the winds but have learned to harness them in a predictable manner, we can harness and—in a very real way—manipulate probability. Early humans' questions have given rise to modern ones: Will the person I'm dating want to get married? Will I make money on my investments? Will I return safely from my trip? Using the power of probability, and the formulas contained in this book, you'll find that answers to these questions can be estimated.

The ancient Greeks made chance incarnate in the goddess Tyche. Later, the Romans called her Fortuna, and she became so wildly popular her temples outnumbered all others. The epigraph above, written sometime in the second century AD, was Fronto's wry comment on her popularity. It is my hope that this book will stand as a temple to both fortune and reason.

The History of Probability Predates History

Probability and chance and the interest human beings have had in them predate historical times. Dice made of animal bones have been found and dated to the Neolithic period, more than six thousand years ago. They look remarkably similar to modern

Knucklebones of sheep or goats (astragaloi) used as dice by ancient Greeks. (Photo courtesy of AbleMedia.com)

dice. In other words, at about the same time the earliest farming societies were being formed, man started playing ur-craps.

These early dice, made of animal bones, are called *astragaloi* (the singular is *astragalus*) and came from particular hoof bones of sheep that had two rounded sides and four square of almost equal size. When early humans played with such primitive dice, starting in prehistoric times and extending to the Greek and Roman eras, the games consisted of betting on the four possible outcomes, disregarding the two rounded sides, since the dice couldn't land on them. Astragaloi remained in use even after the six-sided die, made from bone or carved from wood or stone, was invented, indicating both their usefulness and that there were nostalgists even back then.[1]

In early antiquity, Egyptians and Babylonians played with dice and astragaloi, and so did the Romans. The Etruscans, the mysterious people who lived on the Italian peninsula before the advent of the Romans, played with pentagonal-faced icosahedrons (dice with twelve faces) a thousand years before the birth of Jesus.

According to the Roman historian Suetonius (*Life of the Caesars,* written around AD 100), the emperor Augustus (63 BC–AD 14) was an avid dice player. Suetonius describes the game the emperor loved to play, in which four astragaloi were thrown and the winner was the first person to throw a "Venus," where each of the four four-sided dice showed a different number. Suetonius describes the emperor Claudius (10 BC–AD 54) as having been so consumed with the game of dice that he even wrote a book about dice games. Claudius had a special dice board secured safely to his carriage so that he could play dice while riding around Rome.

Dice games were also popular in ancient China and India. The history and lore of probability are steeped in romantic legend about gambling; one particular tale illustrates this. In the third book of the great Indian epic, the *Mahabharata*, which was written before AD 400, King Rituparna discusses probability and statistics with Nala, a man possessed by the demigod of dicing. Rituparna is described as having the ability to estimate the number of leaves on a tree based on the number of leaves on a randomly chosen branch (a procedure akin to modern statistical methodology). Rituparna says:

> I of dice possess the science
> And in numbers thus am skilled.

The verse suggests Rituparna had some knowledge of probability theory, as his conscious link between dice and numbers would indicate.

The rabbis of the early centuries following the destruction of the Temple in Jerusalem in AD 70 also had some familiarity with probability. This is evidenced in the Talmud, written about the same time as the *Mahabharata*. According to the Talmud, probability arguments were widely used in determining issues related to dietary laws, paternity in cases of adultery, tax distributions, and other issues in which uncertainty played a role. Ancient Jewish texts also inform us that duties by priests in the Temple, while it was still standing, were determined by chance: the priests drew lots for chores, such as cleaning, guard duty, and cooking. According to modern research, Talmud experts seem to have been able to use probability rules for addition and multiplication, and were able to compare the probabilities of different events to make judgments based on relative magnitudes of probabilities.[2]

Surprisingly, the ancient Greek mathematicians, Pythagoras, Euclid, and others, did not spend any time thinking about probability theory. Perhaps they did not view an effort to assess chances as part of mathematics—the main mention of dice in Greek mathematical writings is merely as a tool to help young people learn arithmetic, by adding the dots. There is no discussion of chance.[3]

In the ancient world, both in the East and in the West, dice and astragaloi were used as random devices not only in games of chance, but also for divination. When people desired guidance in making decisions in their daily lives, when military leaders wanted to find out whether the time was right to go to battle, and when kings sought divine counsel on matters of state, they consulted an oracle. The oracles often used the dice to determine the gods' answers to such questions. A "Venus" on the dice would mean "Yes" to the question answered, while a "Dog"—all ones—would mean "No." But there were many other customs and possibilities as well. The use of random devices as sources of divine guidance continued through to the Christian era. There are recorded cases up to our own time of people seeking answers to questions of marriage, jobs, and other issues by means of chance mechanisms.

The basic elements of probability theory as we know it today were formally developed in the seventeenth century by European mathematicians, among them Galileo Galilei (1564–1642), Blaise Pascal (1623–1662), Pierre de Fermat (1601–1665), and Abraham de Moivre (1667–1754). As in India, the European development of the theory of probability was closely associated with gambling, and was motivated by the desire to understand the laws of chance in order to win money while playing against the bank.

Letters to a Young Gambling Addict

The essence of the mathematical theory of probability emerged in France in the 1600s as a result of an unusual partnership between a gambler and a mathematician. The gambler was the Chevalier de Méré, who wanted to find out how to win in the casinos of Europe. The mathematician was none other than the famous philosopher, physicist, and mathematician Blaise Pascal. De Méré came to Pascal and asked him about the probabilities of winning at two different complex games popular in Europe at that time (we will see them later). Pascal wrote to an older mathematician, the famous Pierre de Fermat, and through their correspondence, the mathematical rules of probability were derived. These rules, as expanded through the centuries, are the subject of this book.

CHAPTER ONE

What Is Probability?

PROBABILITY is humanity's attempt to understand the uncertainty of the universe, to define the indefinable. A *probability* is a quantitative measure of the likelihood of a given event. If we are sure that an event will occur, we assign it a one hundred percent probability. If we are sure that an event will not occur, we assign it a probability of zero percent. Other events—those neither certain to occur nor certain not to occur—are assigned probabilities between zero and one hundred percent (or, equivalently, between 0.00 and 1.00, which is the strict mathematical scale of probability). If an event has a probability of 0.5 (that is, fifty percent), then it is as likely as not to happen. An event with probability of 0.1 (ten percent) is not likely to take place; and an event with a probability of 0.9 (ninety percent) will likely take place. Of course, these same numbers can be written as fractions, so that 0.1 becomes 1/10, 0.5 becomes 1/2, and so on.

The figure below is a guide to interpreting probabilities.

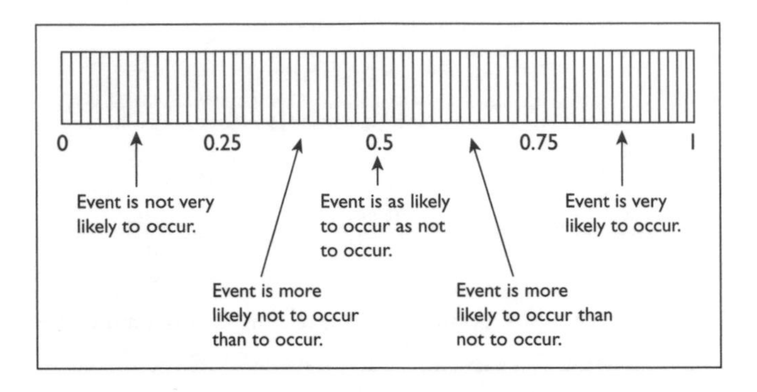

The assignment of a probability, a number between zero and one hundred percent (or 0.00 and 1.00), must be done in accordance with certain laws of logic and mathematics—derived by Pascal and Fermat and others—so that it is valid. We want the probability we assign to an event to be correct; which is to say that, in the long run, if we make many trials of the event (say, rolling a die and noting the outcome) the probability will agree with the results. For example, we assign the number 1/6 to the probability that a die will show a "five." And in fact, in the long run, 1/6th of the rolls of a fair die will show a "five." When early humans rolled that bone die before history began, they noticed that about 1/6th of the time, a "five" came up— or they would have if they weren't too busy running from woolly mammoths. The same held true for the other numbers. But we derive the probability by other means, not by rolling the die a million times. We've come a long way.

You will find an understanding of probability theory is extremely useful. You already use a basic form every day (should you pay for a cab or chance the subway? pull off the highway and get gas now or wait for a cheaper price?); but a deeper understanding will help you make better decisions in business, love, and any other area of life. The probability expert I.J. Good claims that probability theory predates the human race.[4] He argues that animals have a sense of probability—a predator might instinctively assess the probabilities that the prey will choose among various escape routes, and chase down the route that is most probable.

The purpose of this book is to teach you something about the theory of probability, to help you define the indefinable, so that the world around you makes sense and you're able to make better decisions in your life.

CHAPTER TWO

Measuring the Probability

I WANT TO TELL YOU A SECRET: measuring probabilities is as simple as *counting*. Simply count the possibilities of an event and divide this number by the total number of possibilities. (Of course all the possibilities we count must have an equal likelihood—if one possibility is more likely than others, it must be weighed appropriately, but we'll deal with this later.) For example, in a roll of a fair die, there are six equally likely outcomes: one, two, three, four, five, and six. How do we know? Well, it's intuitive, and perhaps the early humans who played with the bone die had the same intuition. The die is perfectly symmetric (or at least ideally it is); it has six sides; so, each of them is equally likely to come up.

> Whenever we have a situation with equal likelihood outcomes, the probability of any outcome is the ratio of the total number of outcomes corresponding to the event, to the total number of outcomes.

For example, what is the probability of rolling an even number? There are three even numbers (two, four, six) out of six equally likely numbers (one, two, three, four, five, six), so the answer is 3/6 = 1/2, or fifty percent. Even odds, as they say in Vegas.

Let's look at cards now. If I have a deck of fifty-two cards that is well-shuffled, what is the probability of drawing an ace? The assumption that the deck is well-shuffled gives us the equal likelihood of outcome for any one card out of the fifty-two cards in the deck.[5] Since we have four aces (the aces of hearts, diamonds, clubs, and spades) out of fifty-two cards equally likely to be chosen, the probability of an ace is 4/52 = 1/13 = 0.0769, or about eight percent. This is demonstrated in the figure below.

The Space of Possibilities for Drawing a Card

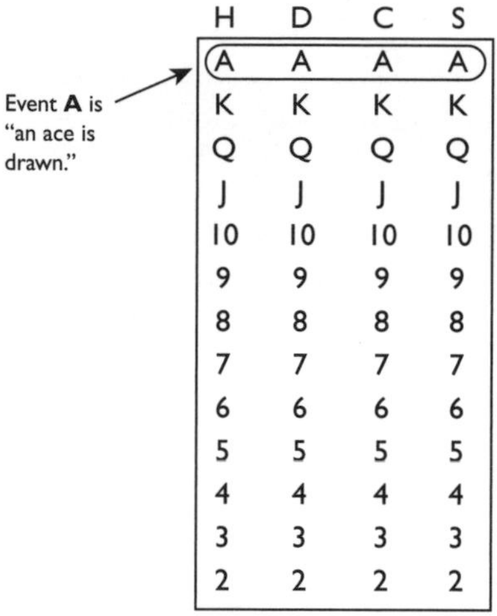

Event **A** is "an ace is drawn."

H	D	C	S
A	A	A	A
K	K	K	K
Q	Q	Q	Q
J	J	J	J
10	10	10	10
9	9	9	9
8	8	8	8
7	7	7	7
6	6	6	6
5	5	5	5
4	4	4	4
3	3	3	3
2	2	2	2

CHAPTER THREE

The Law of Unions

WHEN WE WANT the probability of the occurrence of one of two events, we have a law for finding it, called the *union law.* The union law says the following:

> The probability that at least one of two events will occur is the sum of the probabilities of the two events, minus the probability that both events will occur.

You might find this language tricky at first, but a quick example shows just how simple it is: what is the probability that the card we draw out of a well-shuffled deck is either a heart or an ace (or both)? The answer, by the rule above, is: probability of a heart + probability of an ace – probability of the ace of hearts. The reason for this is that the ace of hearts is *both* a heart *and* an ace, so we need to subtract that possibility (otherwise we will count it twice and overestimate the probability of drawing a heart or an ace).

The figure below demonstrates this rule.

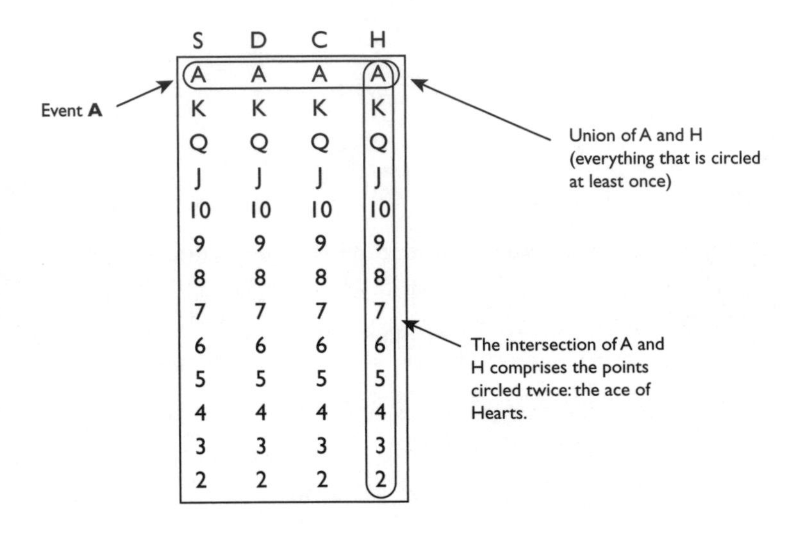

CHAPTER FOUR

Independence of Events

THE ABOVE HAS MOSTLY BEEN SIMPLE AND INTUITIVE. We now come to more realistic situations. Here we encounter one of the most important assumptions in probability theory, the assumption that random processes are—in many situations—memoryless.

If a die turns up "two," does that increase the chances that the next roll will also end up in a "two"? If the die is fair, rather than loaded, the answer is no. In this case, we say the process of rolling the die is *memoryless.* The die does not remember that it just showed a "two," and therefore the chance of a "two" on the next roll remains unchanged at 1/6.

When a process is memoryless, like the roll of a die, successive events are called *independent of each other.* In addition, events that are separated by time or space also tend to be independent of each other. For example, the results of an election in Massachusetts and a rainstorm in Tasmania are independent of each other—if we know that it rained in Tasmania today, that does not increase or decrease the chances that someone will be elected governor of Massachusetts.

> For independent events, the probability of joint occurrence is equal to the product of the probabilities of the separate events. In other words, multiply their two probabilities together, and you get the probability that both events will occur.

By the rule for joint occurrence of independent events, the probability of rolling two "fours" in two rolls of a fair die is the product of the two separate probabilities. Since the probability of a "four" each time is 1/6, the probability of rolling two "fours" in two rolls is $1/6 \times 1/6 = 1/36$. Out of thirty-six rolls, on average you will roll two "fours" once.

The important *multiplication law* of the last section, you will be happy to know, can be extended to events that are not necessarily independent—with one twist. We must multiply the probability of one event by the *conditional* probability of the second event given that the first event has occurred. Huh? Read the example below, and it will all become clear.

> For *dependent* events, the probability of joint occurrence is equal to the product of the probability of the first

event and the probability of the second event *given that* the first event has occured.

For example, suppose that there are ten people in a room—five men and five women. What is the probability that two people chosen at random will both be women?

Solution: We take the probability that the first person will be a woman, 5/10, and multiply it by the probability that the second person chosen will be a woman—but in this case, there are only four women and nine people left to choose from. So, the probability is 4/9. Thus we have: $5/10 \times 4/9 = 20/90$, or, 0.22.

This kind of sampling from a group is called *sampling without replacement.* If we sample with replacement, i.e., allow for the fact that the two choices could result in the same woman being chosen twice, the probability is simply $1/2 \times 1/2 = 1/4$.

WITH THE LITTLE THEORY PRESENTED THUS FAR, we can already develop some interesting applications of probability.

If two dice are thrown, what is the probability that the sum of the dots on the two dice will be four?

Let's assume that one die is white and the other is black.

One way the two dice may fall is the following:

In the case above, we see that the white die shows two dots while the black one shows five dots. We want to build a space of possibilities for the outcomes of two dice, as we did earlier with the cards. How many ways can the white die come up? The set of possibilities is shown below.

Likewise, there are six ways for the black die to fall, as shown below.

By simple logic, there are $6 \times 6 = 36$ ways for a pair of dice to fall. All these ways—the space of all possibilities for the outcome of two dice—are shown on the following page.

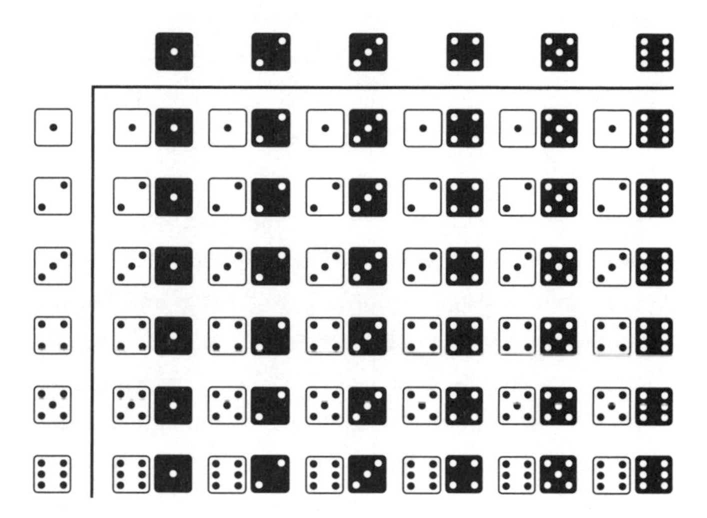

We now look at the space above and count the number of pairs in which the sum is four. We see that the points are: (one, three), (two, two), and (three, one). Thus there are three such ways out of thirty-six total possible ways, and the probability is 3/36 = 0.083, or, roughly eight percent.

If a card is drawn out of a well-shuffled deck of fifty-two cards, what is the probability that the card is either a spade or a "high card" (a face card or an ace)?

Here we use the law of unions: The probability of a spade or high card equals the probability of a spade plus the probability of a high card minus the probability of both a spade and high card. Or: 1/4 + 16/52 - 4/52 = 0.48. Almost fifty percent.

Blackjack

The idea of the space of all possibilities, which we have been discussing as the key tool for evaluating probabilities (and which statisticians call the *sample space of an event*), is only useful when it is available. In gambling situations this is generally true. But in blackjack, for example, the space is not always known exactly. This is especially true since casinos use several decks of cards, so that the gambler has a difficult time keeping track of what cards have or have not come up. What the clever gambler must do is to try to "count" the cards, or keep a rough tally of how many high or low cards are left, and from that tally reach a rough probability that the next card drawn will be helpful or hurtful.[6] Such estimation allows the gambler to decide whether to ask for an additional card ("hit") at any given point in the game. For example, suppose I have been playing and watching the outcomes of the cards for some time, and I now estimate that the pile from which the cards are drawn is rich in low cards. (I've come to this conclusion by observing many high cards, and few low ones, thus far). Suppose further that my hand has twelve points. In this case it will be advantageous, generally, to ask to be "hit," since I can increase my count closer to the winning sum of twenty-one, with a relatively low probability of drawing a face card and going bust.

CHAPTER FIVE

Subjective Probability

UNLESS YOUR INTEREST is exclusively in gambling, this chapter is the most important in this book, for it allows us to apply the techniques of probability to situations that don't obviously call for them.

Situations with No Clear Space of Equally Likely Possibilities

In cases other than games of chance, we rarely encounter a clear set of possibilities that are all equally likely. For example, the set of possibilities {rain, no rain} is generally not a collection of outcomes that are all equally likely. How do we estimate probabilities in such situations? How do we address chance in everyday life?

First, we must define something called an *equivalence lottery*. What is a lottery?

We are all familiar with lotteries, since many states run lotteries in which the winning numbers allow the winner to take away millions of dollars. The probability of winning such lotteries is almost null, and we will discuss these lotteries in more detail later in the book. A simple lottery, for our purposes here, is a collection of balls in a bag. If a bag contains one hundred balls, eighty of which are red, and the chance of winning this lottery is the chance that the ball you draw from this bag is red, then the probability is 80/100 = 0.8, or, eighty percent.

The most important idea of this chapter is to be able—somehow—to "impose" a kind of lottery to events that are not all equally likely. Let's look at our rain example. We have said that the set {rain, no rain} is not a collection of outcomes that are all equally likely—when it comes to the weather, you don't draw from a deck of two cards, one saying "rain" and the other saying "no rain." So, how can we impose a lottery on this situation? Perhaps the day in question is in July, and you know rain is rare in July where you live, and from counting the number of rainy July days over the past several years and dividing by the total number of July days, you get 6/93. All else being equal—meaning that this is an average year, not a drought year or an especially rainy year—you can view rain tomorrow as a draw of a ball from a lottery, with six red balls out of a collection of ninety-three. Now we're getting somewhere. The probability of rain is 6/93 = 0.0645, or, roughly six percent. Of course, we have weather forecasts that use satellite pictures and other methods. But the idea of a lottery will serve us well in cases where no other forecasting is possible.

A Stock Market Example of a Lottery

I want to play the market as a day trader. Al those ads on CNBC have gotten to me, and I want to buy a stock tomorrow morning, with the expectation of selling it in the evening for a profit. I'll buy the stock if I think that the probability of making money on it during the day is good (better than fifty percent). How can I estimate the probability of making money on this trade?

The idea is, again, to try to impose some equal likelihood lottery in a case where one does not overtly exist. What I can do is imagine the stock is drawn from a collection of balls in a bag, out of which some are marked with "up" (the red balls), and the rest with "down" (the other balls in the bag). What are some ways of trying to do this in a "scientific" or logical way? I need to imagine the world of possibilities for this stock. I need to imagine that there are one thousand days identical to this one, and that on each of these one thousand days I have a chance to buy this particular stock. On how many of these days will my stock go up? What we must do is impose a logical lottery on this stock. Suppose that from the economy, the stock market, and the psychology of buyers and traders, I estimate on six hundred and fifty of these hypothetical days the stock goes up.

I have just created a lottery in which the probability of drawing a red ball—i.e., the stock going up—is sixty-five percent. There are some more precise and more scientific ways of imposing the lottery, as we will see.

PROBABILITY CAN BE *objective* or *subjective.* An objective probability is a probability that everyone agrees on. For example, with a fair die, every rational person agrees that the probability of rolling any given number is exactly 1/6. But what about the probability that IBM stock will go up in value tomorrow? This is a subjective probability—subjective because it depends on the person making the assessment. My subjective probability for this event is about sixty percent. Where do I get this number? I looked at a graph of the movements of IBM common stock over the last month and noticed that there was a general upward trend, with some days of decline, and after counting these days, I estimated that the chances of the stock going up tomorrow is about sixty percent. Perhaps your subjective probability for this event is about the same as mine—if you have no better information at your disposal than a mere graph of past performance. If you are a stock market analyst, perhaps you know more; your information set is different (and, one would hope, better) than mine. Perhaps you know much about the computer industry and have noticed a general decline over the short run for similar company stocks. So your subjective probability for IBM's going up tomorrow may be only forty-five percent. Now suppose you are an IBM *insider.* In this case, your knowledge may be vastly better than mine or a stock analyst's. Suppose that you know the earnings report by IBM, which will become public tomorrow, is negative and will disappoint the market. In this case, your subjective probability for the event "IBM stock goes up tomorrow" is perhaps a number that is very small, say, five percent, reflecting the chance that perhaps the stock will go up if this information is already reflected in the stock's valuation. Clearly, the insider's assessment is immensely better than yours or mine. The insider has an advantage over the rest of the market, which is exactly why insider trading must be reported and controlled, and violations of the insider trading rules can lead to jail sentences. When playing the market against you and me, insiders have the equivalent of dice loaded in their favor.

The de Finetti Game

Bruno de Finetti (1906–1985) was an Italian statistician who spent his life developing powerful methods that lie in the middle ground between mathematics and psychology. Incredibly, he found a way to objectively measure subjective probability. (If that sounds weird, keep reading, you'll see what it means.) His brilliant, elegant method is today called the de Finetti Game.

The de Finetti game is a way of making people get in touch with their inner feelings, so to speak; most people lie about probability without even being aware of it—they even lie to themselves. Suppose your friend just took an exam and feels good about how he did on this test. This person might tell you, "I aced it; I'm one hundred percent sure I'll get a perfect score." Now, you know no one believes, truly, that anything (except for death and taxes) has a one hundred percent probability. So the question is, how sure is your friend—*really*—about having aced that test?

The de Finetti game consists of asking your friend a series of questions aimed at assessing his *true subjective probability* for the event of acing the test. Here's how it works:

Tell your friend the following. "Let's play a game. You have a choice. You can either draw a ball from a bag that has ninety-eight red balls and two black balls. If you happen to draw a red ball, I will give you one million dollars. Or you can decide to wait to see how you did on the test, and if you receive a perfect score on that test I will give you one million dollars. What's your choice: draw or wait?"

Presumably, your friend will say, "Draw from the bag." If he doesn't, it means he really has a subjective probability of one hundred percent of acing the test. If your friend's decision is "draw," you ask the next question: "Now there are eighty red balls in the bag and twenty black ones. Do you want to draw, and if you obtain a red ball get a million dollars, or wait to see how you did on the test and get a million if you aced it?" If the answer is now "wait" (for the test result), then we know the subjective probability of acing the test is more than eighty percent but less than ninety-eight percent. So now choose some value in between, such as ninety percent and say: "Now there are ninety red balls and ten black ones in the bag, do you want to draw or wait?" If the answer is draw, try the next question: "There are eighty-five red balls and fifteen black ones, draw or wait?" If the answer is "Draw," try eighty-three red balls. Your friend's answer now might be "I'm indifferent between drawing and waiting." In this case, the subjective probability of acing the test is eighty-three percent. If not, update the numbers of red and black balls in the bag accordingly so that you can bracket the actual subjective probability between a draw and a wait answer.

Play this game, and you'll see just how often people's probabilities change. Interestingly, it's been found that weather forecasters tend to change their probabilities little—because they are used to thinking in term of probabilities. And there's one other caveat: this game tends to fail in matters that are very important to people—for example, love. We can guess this is because the gain of the event in question—say, marrying someone—is more important to many people than the gain from the lottery—one million, in our example. Although not always. . . .

On a historical note, Bruno de Finetti refined his theory of equivalent lotteries by playing the game every week with his students at the University of Rome. They used it to guess the probabilities of a win by different soccer teams in the Italian League.

When an objective probability can be determined, it should be used. (No one would want to use a subjective probability to guess what side a die will land on, for example.) *In other situations, we do our best to assess our subjective probability of the outcome of an event.* Another option in such situations might be to try to buy some information—consult with an expert, for example, a financial analyst. Once our subjective probability is assessed, it can be used in conjunction with the usual probability rules.

Suppose, for example, that your spouse has applied for two jobs within the same company. She feels that there is a sixty percent chance (as assessed through the de Finetti game) she will get job offer *A*; a twenty percent chance for job offer *B*; and a ten percent chance of getting *both* job offers. Using the LAW OF UNIONS, we find that her chances of being offered at least one of the two jobs within this company are: $0.60 + 0.20 - 0.10 = 0.70$, or seventy percent. You can start shopping for champagne.

CHAPTER SIX

The Complement of an Event, and the Union of Independent Events

WHEN THE OPPOSITE OF AN EVENT HAPPENS, we say that the *complement* of the event has taken place. If we are interested in the event that it will rain tomorrow, then the complement of this event is that it does not rain tomorrow.

> Probabilities of all events within a given setting must add up to 1.00, or one hundred percent.

That's just common sense. For example, given that we consider only two possibilities—*Rain* and *No Rain*—we must have the probability of rain plus the probability of no rain equals one hundred percent. There's a one hundred percent chance that it will either rain or not rain tomorrow.

Likewise, if the probability of rain any time tomorrow is forty percent, then clearly there is a sixty percent chance that it will not rain any time tomorrow. We thus have the obvious rule:

The probability that an event will occur, plus the probability that the event will not occur equals one hundred percent. Or, in mathematical terms:

$$P(A) + P(\text{Not } A) = 1.00, \text{ or } 100\%, \text{ for any event } A.$$

And equivalently:

$$P(\text{Not } A) = 1.00 - P(A).$$

Using the above definitions of complements of events, we are now in a position to derive one of the most important and most useful laws in probability theory, the *law of unions of independent events.*

Let's consider two independent events, A and B. If one event or the other (or both) occurred, then it is logically clear that it is *not the case* that both complements of these events occurred. Think about it for a moment. I told you that at least one of A or B happened. So you know that surely (Not A *and* Not B) did not happen. Therefore, the *complement* of (Not A *and* Not B) happened. But the probability of a complement of an event is equal to one minus the probability of the event. Thus:

$$P(A \text{ or } B) = 1 - P(\text{Not } A \text{ and Not } B).$$

But the events A and B are independent of each other, and therefore so are Not A and Not B. We now recall that the probability of the intersection of two independent events is the product of the two separate probabilities. Hence the "and" becomes a simple mathematical product, or multiplication, and we have:

For independent events, A and B:

$$P(\text{A or B}) = 1 - [P(\text{Not A}) \times P(\text{Not B})].$$

The rule extends to the union of any number of events, A, B, C, D, etc., as follows:

For independent events, A, B, C , D . . . we have:

$$P(\text{A or B or C or D} \ldots) =$$
$$1 - [P(\text{Not A}) \times P(\text{Not B}) \times P(\text{Not C}) \times P(\text{Not D}) \ldots]$$

There are endlessly many wonderful examples of the use of the above rules. We take the probability P(A or B or C or D . . .) to mean that *at least one of a number of possible independent events, A, B, C, D, etc., will occur.* The "or" that separates the letters stands for all possibilities (meaning either one of A, B, C, or D, alone will occur, or any pair of such events will occur, or any triple, or all four of them).

A woman has five blind dates with five different men over the next two weeks. The men don't know each other, and the events that any of the dates will be successful and potentially result in a romantic relationship are therefore independent of each other. The woman estimates that the probability of a successful date is twenty percent (she's an optimist). What is the probability that after the five dates she will find at least one suitable man for a relationship? Using the law above:

P(at least one of five independent events will occur) = 1 −[0.8 × 0.8 × 0.8 × 0.8 × 0.8] = 0.67, or sixty-seven percent. You may have noticed that to find the probability of success of at least one date, we subtracted from 1.00 the product of the five probabilities of *no success* for each date, 0.80, in accordance with the law of unions of independent events. If at least one of five independent dates was successful, then it is *not the case* (hence: 1 -) that *all five dates failed.*

Here is another example, this one of a real situation that happened a few years ago. The chancellor of a state university, a man I will call Dr. Charles, was fired and had to look for another job. At a given point in his job search, Dr. Charles was considered for positions at seven different universities around the country. Since none of the jobs were in the same state and the search committees for the positions were separate groups of people not believed to exchange information across universities, Charles made the reasonable assumption that the events of being offered any of these jobs were independent of each other. Let's assume that Charles was applying for jobs at the following universities: Alabama, Barnard, Cornell, Dartmouth, Emory,

Fordham, and George Washington. He was told both by Alabama and Barnard he was a semifinalist. This meant there was a short list of six candidates for the job at each university. At both Cornell and Dartmouth, Charles was at an early stage of his application, a stage in which there were twenty candidates for each position. And at Emory, Fordham, and George Washington, Charles was a finalist: there were three candidates for each of the three positions. Dr. Charles could reasonably assume that if there were several candidates for any given position all candidates were equally likely to get the job. What was his probability of getting at least one job offer? Try to solve this interesting problem on your own. Once you are ready, read on for the solution.

Solution to the Job Hunting Problem:

Since we assume equal likelihood at each application, and the applications are independent of each other, the situation is like throwing seven "dice"—two of which have six equal sides; three of which have three equal sides; and two of which have twenty equal sides. We use the law of unions of independent events, with the first letter of each university denoting the event that Charles gets the job at that particular school, and have:

P(*A or B or C or D or E or F or G*) = 1 – [P(*Not A*) × P(*Not B*) × P(*Not C*) × P(*Not D*) × P(*Not E*) × P(*Not F*) × P(*Not G*)] =

1 – (5/6 × 5/6 × 19/20 × 19/20 × 2/3 × 2/3 × 2/3) = 1 – (72,200/388,800) = 1 – 0.1857 = 0.8143, which is eighty-one percent—not a bad chance at all. Notice that this problem has a nice structure that enables us to multiply all of the top numbers first, and all of the bottom numbers next, and only then do the division, thus making it easier for us and more accurate than, say, trying to multiply 0.6666 by something.

The Gambling Problem of the Chevalier de Méré Explained

Let's return to the story of the Chevalier de Méré (mentioned in the introduction), the seventeenth-century French gambler who contacted the mathematician and philosopher Blaise Pascal to help him solve a gambling problem—and inadvertently inspired modern probability theory.

The Chevalier was confused by the probabilities of two games popular in the French casinos at that time. The first game consisted of rolling a die four times and winning if at least one "ace" (the number one) appeared. The second game consisted of rolling a pair of dice twenty-four times, and winning if a double-ace (the number one appearing on both dice) came up at least once. The benighted knight thought that the probabilities should be equal. He reasoned as follows: "In one roll, I have a one-in-six probability of getting an ace; therefore, in four rolls, the chance of at least one ace is 4 × 1/6 = 2/3. In the second game, my odds of getting a double ace in one roll of a pair of dice is one-in-thirty-six; therefore, in twenty-four rolls of a pair of dice, the chance of at least one double ace is $24 \times 1/36 = 2/3$." And yet in reality, the first game resulted in a win somewhat more often than in the second game. This was known as Paradox of the Chevalier de Méré.

You should see right away that his reasoning is completely false. When the Chevalier contacted Pascal, the famous philosopher, in cooperation with another famous mathematician, Pierre de Fermat, developed our rules above.

Using the rules we have developed in this section, the law of unions of independent events, we get the following.

For the first game: P(at least one ace in four rolls of a die) = 1 – P(no ace in four rolls) = $1 - (5/6)^4 = 0.5177$. This means a win in a little over half the number of games played, on average, in a long sequence of games.

And for the second game: P(at least one double-ace in twenty-four rolls of a pair of dice) = 1 – P(no double-ace in twenty-four rolls of a pair of dice) = $1 - (35/36)^{24} = 0.4914$. This result means, of course, a win in fewer than half the games played, on average, in a long sequence of games.

A very clear example of why the Chevalier de Méré was wrong can be seen with a simple example. If we follow his logic, then in two tosses of a fair coin, the chance of getting "heads" at least once is one hundred percent! Because the chance of heads the first time is 1/2, and therefore the chance of heads at least once in two tosses would be $2 \times 1/2 = 1$. Of course, the correct way to find the probability of at least one head in two tosses of a coin is: $1 - (1/2)^2 = 3/4$.

It is important to understand that you must use the law of unions of independent events (when events are believed independent, match), even if the logic seems to break down somewhat. In their excellent book *Statistics*, David Freedman et al. bring an example of a false use of probability in popular literature. They quote the author Len Deighton as saying in his novel *Bomber* that a World War II pilot had a two percent chance of being shot down on each mission. So in fifty missions, they quote the novelist as saying, the flyer was "mathematically certain" to be shot down. The false reasoning here is, of course: fifty × two percent = one hundred percent. The correct way to solve this problem is (assuming independence of missions): $1 - (0.98)^{50} = 0.64$, or sixty-four percent—which, while relatively high, is quite different from one hundred percent.[7] Adding the fifty separate probabilities of two percent each is wrong because we must "subtract the intersections," which is implicit in the law of unions of independent events. Of course, intersections in this case mean being shot down twice—and that doesn't make any sense. But the rule must be followed nonetheless, or else nonsensical results are obtained.

Interesting Implications of the Law of Unions of Independent Events

Let's play around a little with this powerful law. Here's another problem for your amusement.

A document carrying my original signature (which cannot be faxed) must arrive in California tomorrow morning before 10:30 A.M. for the closure of an important business deal. To maximize my chances that this will happen, I send three such documents, using three different overnight services. From experience, it is known that Service *A* has a ninety percent on-time delivery record. For service *B*, the record is eighty-eight percent on-time performance; and service *C* is known to deliver its shipments on time ninety-two percent of the time. What is the probability that at least one of my documents will arrive on time, and hence that my original signature will be there in California tomorrow before 10:30 as required? The answer may surprise you.

Using the law of unions of independent events we get: $1 - (0.10 \times 0.12 \times 0.08) = 0.99904$, which is 99.9 percent, which is very, very high. What happens if I'm still not happy? I can add, say, two more overnight delivery services, with probabilities of on-time delivery of only eighty-five percent each. What is the probability now that at least one of my five packages will arrive in California before 10:30 A.M. tomorrow?

Again, using the law of unions of independent events, we get: $1 - (0.10 \times 0.12 \times 0.08 \times 0.15 \times 0.15) = 0.999978$. I don't know about you, but for me, 99.9, nine, seven, eight percent is a virtual certainty. I would love such a probability for all the good things in life. So what have we noticed here? We've noticed that *as the number of trials goes up, so does the probability of success.*

You might object, saying: "Well, you *started* with high probabilities anyway, so of course you ended up with a high probability." And your reasoning would seem valid—but it's wrong. Let's look at another example. Suppose that my chance of getting a job is very low because I just bungled up every interview and the job market in my field is awful. Suppose that my probability of getting a job is a mere *half a percent*, that is, 0.005. My claim is that if I can just persevere and apply at a very large number of firms, I can bring my probability of getting at least one job offer to a virtual certainty as well. Let's try for, say, two thousand jobs. If I apply for two thousand jobs, my probability of getting at least one job offer is: $1 - 0.995^{2000} = 0.999956$, or 99.9956%. So even someone with a tiny probability of success a single trial can bring up my probability of success to a virtual certainty by trying enough times (and then, the probability is of getting one *or more* job offers!).

> *With independent trials, keep trying and eventually you will succeed!*

There is a story about a guy who propositions every woman he meets. A friend asks him whether his system works. "Well, I get slapped in the face a lot," he answers, "but every once in a while, I get lucky." The guy may be a cretin, but he knows something about the laws of probability.

Unfortunately, this law works for bad events as well as good ones. If your chances of being hit by a car while crossing a street are some tiny number, cross enough streets in your life and you will be hit. (Of course, crossing streets is not a purely random event, so being careful can lower your probability of being hit.) Eventually, for most people, we don't live long enough to get hit by a car—we die from other causes before we get the chance. All joking aside, however, this also helps explain the increase in prostate cancer in men in America; as they are living longer and not dying from other causes, their odds of developing the otherwise uncommon cancer increase to over fifty percent. Likewise, the experience of the nuclear industry both in the U.S. and in other countries has shown that even with a small probability of an accident at a nuclear power plant on any given day, as the number of days (and months and years) of operation increases, so does the probability of an accident.

Monkeys Typing Hamlet

Carrying the above argument to its limit, anything that can happen (meaning, that has a non-zero chance) will happen—if given enough opportunities. This is how we obtain bizarre results such as the Monkeys Typing *Hamlet* example. There is a very, very small (very close to zero, but not quite zero) probability that a sequence of randomly typed letters will happen to form Shakespeare's *Hamlet*. Actually, we can compute the probability. Let's assume there are thirty kinds of typed characters in English: twenty-six letters, comma, period, semicolon, and colon—we will ignore spaces and capital letters; let's give this monkey a break! The chance that the monkey will type correctly the first letter in *Hamlet*, which is "a" (from: "Act I"), is 1/30. We need to multiply this number by 1/30 for the second character typed to agree with *Hamlet* (meaning, to be the letter "c"). We multiply, rather than use the rule for unions, because we need the first letter to be correct *and* the second one *and* the third one, etc. Continuing the multiplication this way, for each of the one hundred forty-two thousand, nine hundred and forty three characters in the entire tragedy *Hamlet*, we see that the answer is 1/30 raised to the power 142,943. This number is extremely close to zero—but not exactly zero. It is around 0.0... followed by over *two hundred thousand zeros*, and then a one. So if you now set a monkey in front of a typewriter and have the animal type *forever*, dividing the typing into endless sequences of one hundred forty-two thousand, nine hundred and forty three characters each, then you will have created an *infinite* sequence of trials (each comprising of one hundred forty-two thousand, nine hundred and forty three characters) in which we are looking for at least one success, a success being the full play. We use the rule for unions, with each probability being the tiny probability of success in a single "trial"—a trial being typing the full number of characters in *Hamlet* one time. Since the number of such sequences is infinite, it can be shown mathematically that the answer converges to 1.00, or one hundred percent, even though the probability of success at each trial is ridiculously small. Given infinite time (whatever that means), a monkey will type the entire Shakespearean tragedy. And not just that, he'll type it an infinite number of times. And he'll type it an infinite number of times backwards. And in Esperanto. But the concept of infinity is another book entirely.

In 1990, my colleague N. Josephy and I explored what happens if the infinite trials are *not* independent of each other. In particular, we wanted to find out what happens if the probability of success *decreases* from trial to trial, even if that decrease is arbitrarily small. We were motivated in our search by sending a pervious article to a scientific journal that stated on its editorial page, "Rejection rate for submitted manuscripts: 86%." Reading this statement we thought, "Well, if we send this journal ten different articles, what is the probability of at least one of them being accepted?" Of course, we know that the answer is $1 - 0.86^{10} = 0.7787$, or almost seventy-eight percent. This, of course, assumes independence of trials. We also know that if we increase the number of papers sent to the journal, the probability will rise eventually to reach a number close to one hundred percent. With infinitely many submissions to the journal, the probability will *equal* one hundred percent (although even with a finite number of trials we will still get quite close to one hundred percent, as we've seen in earlier examples). What bothered us at the time, however, was the realization that in real life a situation like this was not likely to guarantee independent trials. After our third or fourth rejection an editor would start to dismiss our papers: perhaps more so every time; we'd become the laughingstock of the office. We reasoned that the probability of acceptance might be fourteen percent for a *first* submission but will decrease from submission to submission. We wanted to study this situation theoretically. What happens, we asked, if we have an infinite number of trials (which, if they're independent, equals a one hundred percent chance of success)—but with slightly decreasing probability of success in each trial? Our theoretical analysis led to an interesting finding. We know (for example, in the Monkey Typing *Hamlet* example) that the probability of success on a single trial doesn't matter as long as it is not zero. It doesn't have to be fourteen percent as in this example. Even if it is as small as $10^{-200,000}$ as in the case of the monkey typing *Hamlet*, if we try *infinitely many times*, the probability of an eventual success is one hundred percent. But we found that even with a large starting probability (fourteen percent, or even a larger number such as ninety percent), if that probability decreases by even a tiny amount from trial to trial, the eventual probability of success in infinitely many trials is always *less than one hundred percent.*[8]

But of course, with *independent* trials, as the number of trials increases, so does the chance of success. This is a useful rule to remember, and it has a nice moral: keep trying, and you will succeed! As the next chapter shows, however, in gambling against a casino that is much wealthier than you, there is another rule to follow, and another moral to be had.

CHAPTER SEVEN

Random Walks and the Gambler's Ruin

A RANDOM WALK is a concept in probability theory that takes its name from an imagined real-life situation representing purely random motion: the steps taken by a drunkard stumbling around a lamppost. In one dimension, the drunkard takes a step forward or backward completely at random; his steps are independent of each other. In two dimensions, the drunkard walks left or right or forward or backward, each step random and independent; this way the inebriated person may be found walking anywhere around the lamppost. Hence the term *random walk*.

The random walk represents many situations in real life, among them gambling and the stock market.[9] The simplest example of a random walk is that of repeatedly tossing a fair coin and counting the number of heads that appear, in total, minus the number of tails. If this number is zero (meaning that as many heads have appeared as tails), then we can imagine that

A Random Walk

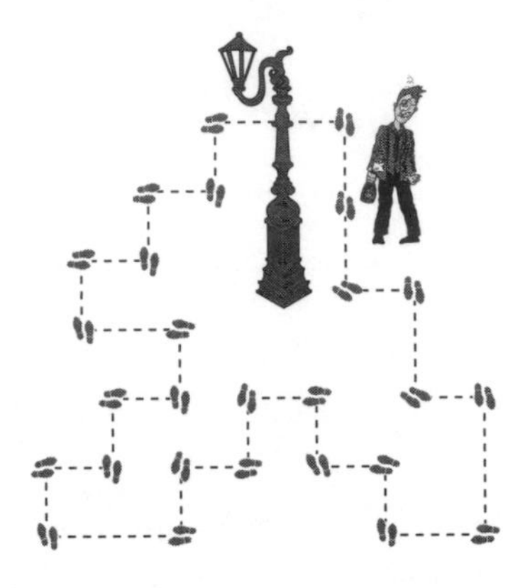

our drunkard—moving in one dimension, forward and backward—is exactly at the lamppost. If the number is plus-two, for example, meaning that by this time there were two more heads than tails counted, then the drunkard is two steps ahead of the lamppost; and vice versa for minus-two, meaning two more tails have appeared so far than heads, and the drunkard is two steps behind the lamppost.

The figure on the following page, reprinted by permission from the classic book by William Feller, *An Introduction to Probability Theory and Its Applications*, shows the result of a computer-simulated sequence of ten thousand tosses of a fair coin.[10]

Notice a very interesting property of the random walk with probability of 1/2 for each of two outcomes. Intuitively, we feel

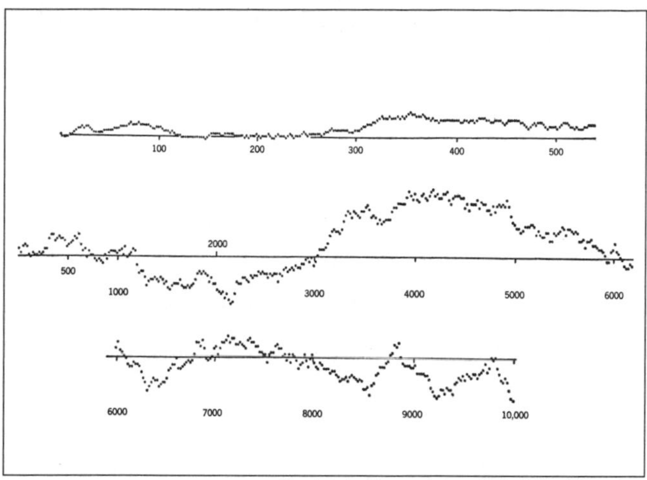

Reprinted with permission of John Wiley & Sons, Inc.

that since the probability of each result, head or tail, is equal, there should be roughly as many heads as tails at any given point. This feeling is indeed true—but *only over the long term.* Short term, there can be many more heads than tails, or vice versa, for relatively long periods, until a crossover of the zero line happens. (Notice the first crossing point is at around one hundred twenty tosses, and that between roughly the point two hundred and fifty and about five hundred and fifty, the number of heads is always greater than the number of tails; and then, between three thousand and six thousand, the number of heads is way over the number of tails and does not come to zero until around toss number six thousand or so!) This is a very surprising finding for most people. This pattern can be proven to be mathematically true in general for the behavior of a random walk. If you think about it for a while, you will realize the strong implications of the behavior of random walk on gambling situations.

The picture above should be enough to convince you that in repeated gambles one does not enjoy the "keep trying and thou shall succeed" moral of the last chapter. The reason for this is that trials aren't free—you have to pay for each gamble. The variable depicted in the graph above is your fortune at each point, the origin (zero) being where you started. When you are below the line, you have a loss; above the line you have won that amount, which is equal to the difference between the number of heads and the number of tails (assuming you win when the coin shows heads). In reality, this process may represent one of the nearly even gambles in roulette: black or red, odd or even; one to eighteen or nineteen to thirty-six. This makes the game even worse for you, because the probability of winning is not 1/2 as in the idealized coin-tossing gamble.

In American casinos there are thirty-six red and black numbers, plus a green zero and double zero. Thus, the chance of winning a single gamble on black or red, odd or even, and one to eighteen or nineteen to thirty-six is equal to 18/38, which is 0.47, or forty-seven percent—less than fifty percent. European and other non-U.S. casinos are a bit better for the roulette player, for they have only one zero. Thus the chance of winning a single gamble in one of these roulette possibilities is 18/37, which is roughly forty-nine percent. (Note that European casinos charge an entrance fee, so that some of the advantage of the gambles is offset.) Typically, you bet an amount of money on, say, the color red. If either a black number appears or a zero or double zero (in a U.S. casino), you lose. The chance of losing is therefore 1 – 0.47 = 0.53, or fifty-three percent. So the more you gamble, the more you lose, on the long run. The picture of the random walk above does not represent you, since this kind of random walk always drifts *downwards*, because the house has a better chance of winning. See the appendix for more on roulette.

But let's suppose that the game is fair, meaning that a coin is tossed and you win on heads (or tails) and the casino wins if the opposite occurs. Equivalently, we can assume a roulette game with no zeros. A famous mathematical theorem states that even in such a fair game, when you play against a much wealthier adversary, like a casino, given enough gamblers, the gambler will lose *with probability 1*—or absolute certainty. This is called the *Gambler's Ruin Theorem.* The mathematics required for proving this theorem are beyond the scope of this book, but looking at the graph above should give you a good idea why it works this way. Notice that while the plot of the data can stray away from zero for long periods of time, it always comes back to zero. Zero is called an *absorbing state* of the random walk.

Bold Play is Optimal

So how should you gamble, or to paraphrase the title of an important book on the mathematics of chance by Lester E. Dubins and Leonard Jimmy Savage, we want to know *How to Gamble if You Must*?[11]

Because the chances are always against you in a casino, most mathematicians and statisticians will advise you not to gamble. Still, people do all sorts of things they're advised not to, and Dubins and Savage—counted among the world's greatest probability experts—addressed in their book a romanticized setting in which a person might gamble. Of course, as we may have guessed, the authors were really interested in deep mathematical theory, but used the gambling example as the setting to develop their results. So here is the story.

You find yourself one night in the middle of the dazzling lights and glitzy activity inside a Las Vegas casino. Having studied the mathematics of chance, you know that in general gambling doesn't pay, and so you have no interest in gambling. But gamble you must. You have twenty thousand dollars on you, but that money is not enough; in fact, on its own, this money is worthless to you. At dawn—and the circumstances here will be left to the imagination—someone will come to get you. If by then you have forty thousand dollars, then everything will be fine; but if you do not have this full amount on you . . . well, let's just say you will be in deep, deep trouble. What should you do?

The question now is not whether to gamble, but rather *how*

to gamble. You must gamble, so the facts that in general gambling will result in your losing money, and that chance works against you, do not enter here. What you want to do is find a way to maximize your chances of having forty thousand dollars by dawn. Of all the bad gambles you can make (bad because they generally reduce your wealth), you need to find the strategy that will be the "least bad one"—the strategy that will give you the highest probability of making your goal.

There are two extreme strategies here, and the rest of them lie in between these two extreme cases. The first strategy, obviously, is to go to the nearest roulette table (the story here is limited to roulette), place your entire twenty thousand dollars on black, close your eyes, and pray...and, if you win, stop gambling immediately and wait for your rendezvous at dawn. What is the probability of being able to come up with the required forty thousand dollars if you follow this strategy? Clearly you will have gambled only once, and we know that (in Las Vegas casino, or any other U.S. casino) the probability of a win is 18/38, which is forty-seven percent. We will call this strategy *bold play*, since you boldly placed your entire twenty thousand dollars at the mercy of chance, rather than keeping some of it in reserve.

The strategy at the other extreme is called *cautious play*. In this strategy (you know you have an entire night to gamble), you bet the minimum at each gamble: each time you place only one dollar on black or red. What is the probability of coming up with forty thousand dollars by dawn using this strategy?

In their book, Dubins and Savage derive the probability formula for this kind of play. The probability of reaching a fortune of size $m + n$, starting with a fortune of size m by playing a series of bets with one unit at stake every time, is given by P. In the formula, p stands for the probability of winning a single bet, and q, which is $1-p$, is the probability of losing a single bet. The formula for the probability of reaching your goal of $m + n$ dollars is:

$$P = \frac{1 - (q/p)^{m}}{1 - (q/p)^{m+n}}$$

In this example, $m = 20{,}000$, and $n = 20{,}000$; we have that $p = 18/38$, and $q = 20/38$. Therefore,

$$P = \frac{1 - (20/18)^{20{,}000}}{1 - (20/18)^{40{,}000}}$$

Since this is very difficult to compute, let's look at an intermediate strategy. Suppose that we write the numbers in the formula above in thousands, and that each gamble is one thousand dollars. Then the probability of reaching the goal of forty thousand dollars by gambling a thousand dollars at a time is given by:

$$\frac{1-(20/18)^{20}}{P=1-(20/18)^{40}}=0.11$$

The probability is therefore only eleven percent, which is *less than a quarter* of the probability of winning using bold play. If you want, you can use a computer and evaluate the probability of betting one dollar at a time—you will find that it is much smaller than eleven percent. So any way you look at it, bold play is optimal.

There is an intuitive reason for the fact that bold play is optimal. Since the laws of probability generally work against you in casino games (I say *generally* because if you can count cards accurately in blackjack, perhaps you will have an advantage; but most casinos nowadays make this task very difficult for the gambler), the strategy that minimizes your exposure to the unfair advantage of the house is the optimal one. In other words, the fewer games you play, the better off you are.

What about doubling your bets when you lose?

There is a mathematical property of sequences and series of numbers that proves that for any positive integer n, the sum of the numbers $1 + 2 + 4 + 8 + \ldots + n = 2n - 1$. Try it: $1 + 2 = 2(2) - 1$; $1 + 2 + 4 = 2(4) - 1$; $1 + 2 + 4 + 8 = 2(8) - 1$; and so on. When people discovered this mathematical quirk, very long ago, they saw in it a "fail-safe" method of gambling: double your bet every time you lose and eventually when you win, you'll be one chip up!

An example shows this. Suppose I bet a dollar. If I lose, I will now bet two dollars (double the initial dollar). Now, if I win, I'll have four dollars, but I'll have invested only three dollars: one on my first bet and two on the second. Suppose I lost three times in a row, doubling my bets every time. Then my investment is 1 + 2 + 4 = 7. I now double again, placing a bet of eight dollars. If I win, I'll have sixteen dollars, but I've only invested 1 + 2 + 4 + 8 = $15, so I'm still one dollar ahead!

But nothing in a casino is fail-safe, or else they'd all be out of business and converted into retirement homes for aging mathematicians (or something). The problem with this system of bets is there's a chance the sequence of losses you incur will be so extensive you'll lose all your money and not be able to continue betting. And when that happens, your loss is liable to be enormous; doubling your money adds up fast—by the tenth loss, assuming you're using ten-dollar chips, you'll be over ten thousand dollars in the hole. How comfortable will you feel placing twenty thousand, forty-eight dollars on a single spin of the wheel? And before you say "Ten straight losses, that'll *never* happen," remember what we learned from the monkeys typing *Hamlet*: given enough iterations, everything happens. That's why the doubling strategy is a sure-fire loser, and why Las Vegas, Monte Carlo, and Atlantic City are still in business.

But...I must admit that I have followed the doubling system and made small amounts of money. As the Bold Play paradigm might have taught you, the secret to successful betting is knowing when to stop. About a dozen years ago, my wife, Debra, and I vacationed on the island of Malta in the Mediterranean. Malta has an elegant casino built on a spit of land jutting into the sea, and every evening we went to the casino. I placed bets in multiples of about twenty dollars in local currency. *As soon as I won once, we left the casino.* If I lost my first bet, I placed forty dollars; and if I lost this amount, eighty dollars. Over the week we were there, I was never out more than one hundred and sixty dollars at any given time. And every night, as soon as I won and was twenty dollars ahead, we left the casino never to return to it that evening. We then went to a very nice restaurant overlooking the harbor and had a wonderful dinner paid for by the evening's winnings. If you play small amounts, and limit your exposure to risk, you might win. But we were reasonably lucky.

AS WE MENTIONED EARLIER, random walks can model the stock market as well. The movement of a stock up or down is given to the vicissitudes of chance. Of course, economic theory tells us that the values of the shares of publicly traded companies should grow as companies make money and the economy grows. This is generally true, but over short periods, even several years at a time—as we've seen with the Dow Jones since 2000—the economy can perform less well than expected and the markets for publicly traded securities decrease. At any rate, chance is a major player in the markets, so even during periods of market growth, the daily movements, the hourly movements, and the instantaneous movements of the prices of stocks are random. Since the amounts of increase or decrease are not necessarily one dollar (or one whole unit) at a time, but can be of smaller or greater amounts (as well as for other reasons having to do with trades taking place over a continuum of time rather than at discrete intervals), the random walk assumption has been refined to that of a physical process called *Brownian motion.* The Brownian motion model has been borrowed from physics, where it describes the motions of dust or smoke particles in the air as they're constantly hit by air molecules and recoil in a random direction by a random amount. A graph of the movements of a stock over an hour, a day, or a week, will reveal to you this pattern. Complex mathematical models that exploit this description of stock movements, the most successful of which is the Black-Scholes Option Pricing Model, have been used with some success. The random walk, with or without an upward drift describing the general growth of the economy, and the Brownian motion models are models of randomness. In the next chapter we will explore the idea of randomness and test its meaning.

CHAPTER EIGHT

What Is Randomness?

YOU WALK INTO A BAR, and you notice the occupancy of the twenty stools. Denoting an occupied seat by S and an empty seat by e, you see the following:

SeSeSeSeSeSeSeSeSeSe

Do you think that the people at the bar chose their seats at random?[12]

Most people would say No. The reason is common sense: everyone can see that there is an empty seat between every two people. This is clearly an unfriendly collection of people, none of whom want to sit next to another person.

Suppose, on the other hand, that the same ten people at the bar occupy their seats as shown below.

eeeSSSSSSSSSSeeeeeee

Would anyone believe that these people chose their seats purely at random?

Here again the logical answer is No. This is clearly a group of friends who came here together—or maybe they are all huddled together in front of a television set watching a ballgame.

What about the next case:

SeeeSSeeSSSeSSeeeSSe

It is possible, in this case, that the seats were chosen *purely at random*, that is, without looking whether you sit next to someone or not. But it's just as likely that you have here a lone person, three pairs of friends, and three friends sitting together.

It is important to note that the first arrangement of twenty elements of two kinds (S and e) could possibly have arisen by chance, and so could the second one. But the *probability* that the first or second arrangements are random is very small. For the third arrangement, pure randomness is much more likely. How can we tell when something is random and when it is not? There is a statistical test that can make a determination of randomness, with a given probability that the statement is correct. The test can be found in statistics books, and it is called the *runs test for randomness.*[13] In the first case above, in which each person is separated by an empty seat from the next one, the test leads us to believe that the arrangement is *not* random. The probability that a sequence like this one arose purely by chance is less than 0.001, meaning less than one tenth of one percent! The same is true in the second case, in which all of the people are sitting together. In the third case, the assumption of randomness cannot be rejected, unless one is willing to take a very high probability of being wrong (over eighty percent). These numerical probabilities come from statistical tables that can be found in books on practical statistical inference. The important point for us here is that arrangements that do not look random are probably not, and vice versa. Incidentally, the runs test for randomness can be used for determining whether random numbers generated by various mechanisms truly are random. In such tests, the digits are labeled as odd or even, giving us exactly the situation as above, in which elements are classified into two groups.

Why Do Good Things Come in Threes?

As we see from the sequences above—and from the graph of the random walk in the previous chapter—randomness does not tend to result in any order. In the case of the random walk, a coin that has a fifty percent chance of falling heads or tails does not often fall in an orderly HTHTHT; but rather, as the graph of the sum of heads minus the sum of tails shows, can stray off and return to the zero line only after long—even *very long*—periods of time. Likewise, in the example above of a random arrangement of items of two kinds, it is very unlikely the elements will appear by chance as SeSeSeSe…, etc. What we have is the following:

> Pure randomness leads to partial (and often unexpected) aggregation.

Look at the third case of the bar stool example above. Here we have the arrangement:

SeeeSSeeSSSeSSeeeSSe

Suppose that over twenty days, ten good and ten bad things will happen to someone. It is very unlikely that these things will alternate perfectly: Good thing, bad thing, good thing, bad thing, etc. It is much more likely, by the rules of chance, that some—though never perfect—aggregation will take place. If good things are symbolized by an "e," then the arrangement above gives us a good pictorial explanation for the maxim good things happen in threes.

Of course, there are other reasons for the folk saying. Once someone has three good things happen in a row, he or she takes notice of it: "See, good things happen in threes!" (rather than wait, and perhaps a fourth good thing will happen, proving in this particular case that good things really happen in fours not threes). And probably no one takes notice when good things happen in twos rather than threes, or when bad things happen in threes. The point, however, is there is a probability theory reason for *some kind of aggregation* of good or bad things, and that the saying about good things coming in threes is a reflection of this statistical phenomenon. In chapter eleven we will explain another phenomenon noticed by people—the persistence of luck, either good or bad.

CHAPTER NINE

Pascal's Triangle

THE GREAT FRENCH philosopher and mathematician Blaise Pascal, who in the seventeenth century helped the Chevalier de Méré understand something about the probabilities in gambling, and devised with Pierre de Fermat the basic elements of modern probability theory, made an interesting discovery. Today, it's called Pascal's Triangle, and it looks like this:

```
                1
              1   1
            1   2   1
          1   3   3   1
        1   4   6   4   1
      1   5   10  10  5   1
     .   .   .   .   .   .   .
   .   .   .   .   .   .   .   .
```

As you can see, the triangle is constructed from the principle that every number in it is the sum of the two numbers placed above the number. Where no numbers lie, we assume the value is zero. For economy of space, we ended it after six lines, but it is of course, infinite. The triangle has many interesting properties, some dealing with prime numbers and other applications in pure mathematics; and in the theory of probability it can be used for obtaining the coefficients of what is known as the "binomial distribution." But an even simpler application of the famous triangle is what will interest us here.

Pascal's triangle gives us a way of computing the probabilities of the number of heads or tails in any number of coin tosses (or the number of boys and girls in any number of children born to a couple; or any of similar equal-probability phenomena). Let us see how this works.

We ignore the top "1" in the triangle, since it is only used to seed the rest of the numbers. Looking at the second row of the triangle we see the numbers 1 and 1. This represents for us the fact that if you toss a coin once, we will have a 50-50 chance of getting a head or a tail. The next row of the triangle—1 2 1—represents the fact that if we toss a coin twice we have one chance in four of getting no heads at all, two chances in four of getting one head, and one chance in four of getting two heads (of course, the same is true of tails). By the same reasoning we get from the next row the result that in three tosses of a fair coin we have one chance in eight of getting no heads, three chances in eight of getting one head, three chances in eight of getting two heads, and one chance in eight of getting three heads (again, the same holds true for tails). The last row in the triangle tells us if we toss the coin five times, there is one chance in thirty-two of getting no heads, five in thirty-two of getting one head, ten in thirty-two of getting two heads, ten in thirty-two of getting three heads, five in thirty-two of getting four heads, and one in thirty-two of getting five heads. If the chances of a boy and a girl are equal, then by Pascal's triangle only one of every thirty-two families with five children will have all girls or all boys.

Notice an interesting property of probabilities that is apparent from inspecting the numbers in Pascal's magic triangle: as the number of trials (tosses of the coin) increases, the probability of an even split becomes smaller! Thus the probability of getting half heads and half tails (or half sons or half daughters) declines. In particular, from the row-before-last we see that in four tosses, the probability of two heads and two tails is only 6/16. Add a row to the table and you will see that in six tosses of a fair coin there is only a 20/64 chance of an equal number of heads and tails. Try this with larger numbers of tosses. If you toss a coin twelve times, the probability of an even split of heads and tails is as small as twenty-three percent.

CHAPTER TEN

The Inspection Paradox

THE INSPECTION PARADOX is one of my favorite topics in probability theory. Let's start with good luck. When you buy a light bulb, you read on the cardboard package "Average life, 2,000 hours."[14] But you *know* that the same light bulb right now in the lamp above your desk has been burning for over a year. The stated average of two thousand hours, at eight hours a day, computes to two hundred and fifty days. And you've had this light on for eight hours a day for an entire *year*, which is much more than two hundred and fifty days. What's going on here? Do all light bulb manufacturers habitually under-estimate the average number of hours their light bulbs last?

And now for the persistence of bad luck. The bus company tells you that Bus no. 57, which stops a block from your house, runs every ten minutes. "So, when I go down to the bus stop," you reason, "I arrive—on average—in the middle of the ten-minute interval between busses." That's a reasonable assumption.

So, by this logic, you should wait, on average, exactly five minutes until the bus comes. But you *know* that you almost never wait just five minutes for the bus. We all wait longer than expected. Sometimes you wait even twenty minutes for that bus! What's going on here? Why do we always have bad luck with waiting for busses?

And then there's the flashlight you keep in your garage. You probably didn't notice, but when you bought the batteries for this flashlight, the package said these batteries will last a total of three hours on average. Now you *know* that the total number of hours you've used this flashlight is way over three hours. What's going on here?

Finally, here is an example you may not have expected at all. Do you know the average longevity for a person of your gender and country? (In the U.S., it's seventy-eight years for women and seventy-four for men). The chances are *better than even* that you will live longer than that average.

THE ABOVE EXAMPLES are all manifestations of the *Inspection Paradox.* This paradox arises when we *inspect* some random process. In the first example, you inspect the light bulb. Before an inspection, meaning before the light bulb is used, the average total lifetime of the light bulb is some number—in this case, two thousand hours. But let's think a minute: this stated average incorporates the probabilities that the light bulb will fail in its first hour of operation (plus its first week, its first month, etc.). But your light bulb—the one that's in right now—has already endured, surviving its first day, first week, first month. It can no longer fail in any earlier time than the present; therefore its total expected lifetime is *longer* than that of a light bulb you just bought. But you say: "What's the difference between the light bulb I buy now and one that's in my lamp at this time?" There *is* a difference. The difference is the inspection: you inspect the light bulb that's in your lamp. That light bulb has *lived* already and hence can no longer die at an age younger than it is right now. The same holds for batteries—and people's lives. A person alive today will live, on average, longer than the expected longevity for his or her gender or nationality or ethnicity or any other category. (Note what we're discussing has nothing to do with the increase in people's lifespans brought about by advancements in medicine or technology—in fact, you could live in a society of *declining* lifespans and still live, on average, longer than the expected longevity.) The reason for this seeming paradox is that the expected life of a person or thing that has already begun its life—e.g., you—is greater than that of something yet to be born, and still susceptible to death or failure from a cause you've already survived. For example, someone alive today can no longer die from infant mortality; and a person who is ninety years old can no longer die at age seventy-five. But the best way to understand the inspection paradox is by looking at the bus-waiting example.

Suppose that busses arrive every ten minutes, on average. But sometimes the bus takes longer to arrive, and sometime less. The average is ten minutes. Let's graph, below, some of these interarrival intervals of your bus.

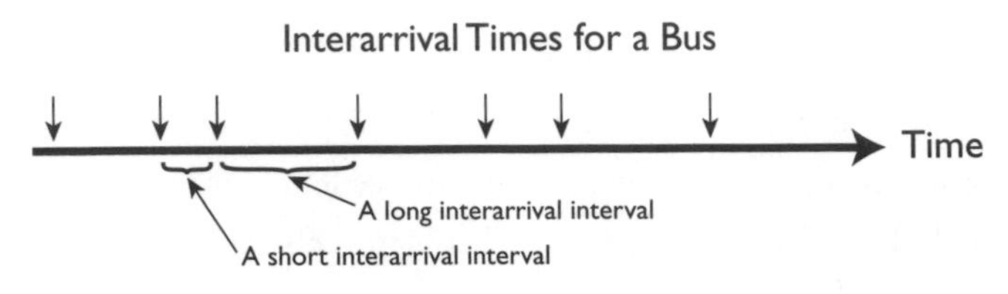

You can see from the figure above that some of the intervals between two bus arrivals are long and some short, but the average interval is still ten minutes. Now, let's say you arrive at the bus stop in the midst of one such interval; as you can see, you are much more likely to land in the middle of a *long* interarrival interval than in the middle of a *short* one. And hence our paradox. The average wait for you—before you go to the bus station—is indeed five minutes; but because you are more likely to arrive during a long interval than a short one, you tend to wait longer than expected.

THE INSPECTION PARADOX manifests itself in other forms as well, and we had better be prepared to look out for the mischief it can wreak. Several years ago, a research on longevity concluded that the longevity of men was highest in Japan, followed by Israel, where a man's average lifespan was reported to be 76.7 years. I grew up in Israel, and this statistic didn't quite make sense to me when I saw it. Israel, with its many chain-smoking, pot-bellied men was second in the world for longevity of men? Something was wrong. The culprit was the inspection paradox.

Mathematically, what the inspection paradox says is that a probability distribution of a quantity that has already started its life is shifted, leading to a larger average than would otherwise be expected. Hence, when you go down to your bus station, the distribution of bus interarrival times has already started. This would not be true if you arrived just as a bus has come and gone—then the interarrival distribution would be starting afresh. But since the bus has been gone for a while at the moment you arrive at the bus stop, the distribution has already begun, and (as can be proved mathematically) your *expected* (or average) wait plus time since the last bus left is longer than the average between-busses interval. Likewise, with longevity, the average longevity should be the average expected lifespan of a person from *birth to death*. In a static population, it's a correct reflection of how long the average man or woman can expect to live. Japan is a reasonable example of a country with such a population. But Israel is not.

Israel is a country whose population consists of a large percentage of immigrants. These immigrants skew the longevity statistic upwards. The reason is simple: an immigrant, arriving at his or her new country at a certain age, can no longer die at any age younger than the present. Let's look at an extreme case. Suppose you have a country where everyone is an immigrant, and everyone arrives in this country at age eighty. Clearly such a country will have an inflated longevity—some age higher than eighty. But even if all immigrants arrive at age six months, they still raise the longevity artificially, since they can no longer die from infant mortality. The life distribution of an immigrant is like the life distribution of waiting times for a bus: once it has started, it is longer, on average, than that of a person who is just born. When a large percentage of the population of a country is an immigrant, that country has an upwardly inflated longevity.

CHAPTER ELEVEN

The Birthday Problem

ONE OF THE MOST SURPRISING RESULTS in probability theory is that sometimes we think an event should be rare, while in reality it is much more probable. We've seen some examples of this observation in our consideration of random walks, for example.

But the most striking example of this is the celebrated *birthday problem.* Here is something you can test at home. Get a collection of people in a room and ask them their birthdays. A priori, you would think that finding two people who share a birthday is hard. There are three hundred and sixty-five days in a year, and it would seem that with so many possibilities for any birthday, there shouldn't be much aggregation. Once the first person "chooses" a day, the second person still has three hundred and sixty-four choices for a birthday that will not coincide with that of the first person. And the third person can "choose" three hundred and sixty-three days without clashing with the previous two; and so on. It seems that with three hundred and sixty-five "boxes" for balls corresponding to people to fall into, there are too many choices for a box to contain two or more balls.

Balls Falling Into Boxes

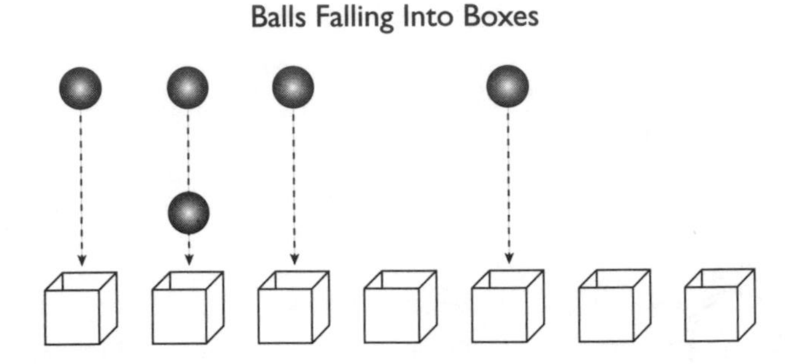

But surprisingly, nature doesn't work this way—things are not at all as they seem. Have a party with twenty or more people, and if you do this every week, you will find that about *half the time* you have a match: two people with the same birthday! Try it.

Why does chance work this way? Well, let's work out the probability. You know already a few things about how to compute probabilities. The probability we seek is that of the union of independent events, because the birthdays of different people are independent of each other. You know that the probability of *at least one match* of people's birthdays is equal to one minus the probability of *no match*. Now, the probability that the second selected person will not match the birthday of the first is 364/365 (there are three hundred and sixty-four empty boxes once one box has been filled by the first person you choose to "inspect"). Now, the probability that the third person you select will not match the previous two is 363/365 (for, if the first two didn't match then there are three hundred and sixty-three empty boxes and two full ones); and so on. It works out that with exactly twenty-three people in the room, the probability of at least one birthday match is . . . well, let's see:

$$P(\text{At least one birthday match}) = 1 - (364/365 \times 363/365 \times 362/365 \times 361/365 \times 360/365 \times 359/365 \times 358/365 \times 357/365 \times 356/365 \times 355/365 \times 354/365 \times 353/365 \times 352/365 \times 351/365 \times 350/365 \times 349/365 \times 348/365 \times 347/365 \times 346/365 \times 345/365 \times 344/365 \times 343/365 = 0.5073,$$

which is over a fifty percent probability!

So with only twenty-three people in a room, and three hundred and sixty-five days in a year, there is a better than even chance that two people will share a birthday. This is a remarkable finding, and it continues to amaze people ever since it was discovered many decades ago. Even more surprising, when fifty-six people are present in a room, there is a ninety-nine percent probability that at least two of them share a birthday! How can we get so close to certainty when there are only fifty-six people and a total of three hundred and sixty-five possible days of the year? Chance does seem to work in mysterious ways. If you have three hundred and sixty-five open boxes onto which fifty-six balls are randomly dropped, there is a ninety-nine percent chance that there will be at least two balls in at least one of the boxes. Why does this happen? No one really has an intuition for such things. The natural inclination is to think that because there are over three hundred empty boxes left over after fifty-six balls are dropped, no two balls can share the same spot. The mathematics tell us otherwise, and reality follows the mathematics. In nature, we find much more aggregation—due to pure randomness, rather than despite of it—than we might otherwise expect.

Mathematicians have developed formulas for estimating the number of "balls" necessary for even odds (fifty percent probability) and for ninety-five percent probability of at least one match, given any number of categories, or "boxes."[15] These estimates are very simple to compute and are a good substitute for calculating the products we used earlier.

> 1.2 times the square root of the number of categories gives the number of "balls" required for even odds that at least two share some characteristic.

And:

> 1.6 times the square root of the number of categories gives the number of "balls" required for ninety-five percent probability that at least two share some characteristic.

Thus, $1.2 \times (\sqrt{365}) = 22.926$, which agrees with what we've computed directly earlier, namely that with twenty-three people there is over a fifty percent probability of at least one shared birthday.

And for a ninety-five percent probability of at least one shared birthday, we have $1.6 \times (\sqrt{365}) = 30.57$. So that if you have thirty-one people in a room, there is a ninety-five percent probability of at least one shared birthday.

The formulas above are general and can be used for any number of categories (here, these categories or boxes stand for the days of the year). Suppose we want to know the number of people needed for giving us a fifty percent chance that two people will share a birth-week, meaning have their birthdays in the same week. The number of categories here is fifty-two, since there are fifty-two weeks in a year. For even odds we have $1.2 \times (\sqrt{52}) = 8.65$. Thus if nine people are gathered together, there are even odds that at least two of them were born in the same week of the year.

The categories (the boxes into which the balls fall) could be anything. Suppose that there are ten thousand categories. For example, these could be combinations of your birthday and your profession (if there are twenty-eight professions; since twenty-eight times three hundred and sixty-five is a little over ten thousand). This is an immense number of possibilities. The question is, how many people gathered in a room would give us even odds, and how many would give us ninety-five percent probability, of finding a match?

$1.2 \times (\sqrt{10{,}000}) = 120$ for the fifty percent probability of a match;

and

$1.6 \times (\sqrt{10{,}000}) = 160$ for a ninety-five percent probability of a match.

These results never cease to amaze me. If there are ten thousand possibilities, and you gather only one hundred and sixty people in a room—that is only 1.6 percent of the total number of possibilities—you will have a ninety-five percent probability of at least one match. Mathematically, of course, we see that as the number of possibilities increases, the number of required people for a fifty percent or ninety-five percent probability of a match increases more slowly—as the square root of the number of categories.

The birthday problem with all its attendant expansions is a problem of *aggregated* coincidences. If we say that among a group of twenty-three people there is a fifty percent chance that at least one pair of people share a birthday, we do not say *which* pair. And, in fact, this is the reason why the probability is so high. Aggregation has many possible ways of taking place. We do not specify which box will contain two balls, only that there is a high probability that at least *one* of the boxes will contain two or more balls. If, for example, you wanted to know what the probability is that in a group of twenty-three people, including yourself, at least one person will match *your* birthday, that probability—because it specifies you rather than being open to any two people—will be much smaller than fifty percent. Let's compute it:

We know how to do this one: it's the probability of the union of independent events.

P (at least one person of the remaining twenty-two people in the room matches my birthday = $1 - P$ (everyone in this group "misses" my birthday) = $1 - (364/365)^{22} = 1 - 0.94 = 0.06$, or only six percent! With one hundred people in the room, the probability that one will match your birthday is only twenty-four percent. And with three hundred and sixty-five people in the room, that probability is *still* only sixty-three percent. The probability is about the same if you have three hundred and sixty-six people in the room; still only sixty-three percent. Note, however, that with three hundred and sixty-six people in the room, the probability of a match of *any* two people—meaning the original birthday problem—is one hundred percent.

CHAPTER TWELVE

Coincidences

WHEN DISCUSSING THE PROBABILITY that someone will match your birthday, we are entering the realm of coincidences—those striking random events that excite us when they happen in our everyday life. Everyone has them. "What do they mean?" we ask ourselves. In fact, people look for coincidences. Are you a Libra? Are you also from Denver? We often ask such questions, and indeed, part of the reason for them is to find something in common; but another is the power or mysticism we attach to coincidences. Everybody has a good story about one.

The most striking coincidence in my life happened some eight years ago when I flew from my home in Boston to Chicago to meet Scott Isenberg, the new editor assigned to my statistics textbook. We were sitting in a restaurant talking about the book, and also getting to know each other.

"I'm from California," Scott said.

"My wife is, too," I said.

This, of course, was a trivial fact, since there are thirty million people living in California. We continued talking about other things; after a while, Scott said, "When I was growing up in California, there were orange groves there . . . now there are just condos."

"That's interesting," I said, "my wife talks about those orange groves, too."

He nodded and the subject changed again. A little later, he said: "This is quite different from how things used to be in my high school . . ."

"Funny," I said, "my wife says the same about her high school."

He looked at me and asked, "What high school did she go to?"

"Pomona High."

"Me too," he said.

"What year did she graduate?"

I told him and he said "Me too," and then asked: "What's her name?"

I said "Debra," and I told him her maiden name.

"I know her!" he said. As it turned out, they were very good friends throughout high school. After this amazing coincidence, Scott became not only my editor, but a good friend, and renewed his friendship with Debra after twenty years.

A coincidence such as this one is a rare event. And so are other similar stories—many of which happen on airplanes. My friend Scott Petrack recently flew from Paris to Boston. Soon after take-off, he started a long conversation with the person sitting next to him (he's a personable guy and loves to talk). Before they landed, the two of them discovered they'd sung in the same choir in Tanglewood twenty years ago. A week later, Scott received a business-related phone call from an unknown person, and in the course of conversation discovered that this person, too, had sung with him in the same choir in Tanglewood. Not once in the intervening years had this happened to him.

These stories, while amazing, and certainly constituting relatively rare events, happen to all of us who fly on airplanes or in other ways meet large numbers of strangers. In my own life, there have been enough such coincidences (mostly on airplanes) that I could fill a book with them. Naturally, I became very intrigued by them and wanted to study them further. How rare are such events?

It turns out a mechanism similar to that of the birthday problem rules coincidences of the everyday type discussed above. Here we have only one person (you), rather than a collection of twenty-three or fifty or two hundred people among whom a coincidence can take place. Because in the earlier cases we did not specify exactly which pair of people would have a coincidence, the probability of such a coincidence rose dramatically. But once we specify *you* as the one person to have a coincidence, the probability becomes very, very small (as we saw with the example of someone sharing your birthday. It is the unseen yet natural aggregation that takes place in random phenomena that causes the probability of at least one, unspecified shared birthday, to be much higher than expected.

Well, in the specific case of coincidences we are concerned with here, there is an *unspecified* element, which makes the probability of coincidences higher than we otherwise might have thought. Although the person is specific (again, you), we do not specify *what kind of coincidence* will take place. In the example of my editor and my wife, such a coincidence could have been anything: Scott could have gone to school with my sister or my neighbor or one of my professors in college or one of my friends. And the coincidence could have involved someone other than Scott; it could have been one of twenty other people I met during my visit to Chicago or someone in line to board the plane or during another trip altogether. I'm not saying this coincidence is common—only that it's not as rare as we think. The reason is that the aggregation of causes happens in so many possible ways that eventually a coincidence becomes unavoidable.

But the topic of coincidences is even richer than that. There are many kinds of elements that come into play. Let's consider the coincidence of having something in common with the stranger sitting next to you on an airplane. What will a careful analysis reveal?

The population of people who travel on airplanes on a regular or semi-regular basis is not that large. Who are they? Professors, lawyers, doctors, researchers, government officials, business people. Thus, we are talking about a smaller group than the entire population of the country. Within this self-selected group, there are networks and connections that go far beyond what meets the eye: for example, professors tend to know not only other professors, but also government officials, through academic research supported or connected with the government; business people know not only other business people, but lawyers, because they so often work together; and so on. These "hidden," built-in connections make the possibility for a coincidence higher than we might expect. Making a reasonable assumption about the average number of people a member of this population knows, as well as other assumptions, I have come up with an estimate of a 1.5 percent probability that you will have a coincidence with your seat mate on a given flight. I define a coincidence as two strangers knowing one person in common. Note there's an information barrier to pass here. If you sit quietly on the plane and just say things like, "Would you kindly move so I can go to the bathroom," then you'll never discover your seat mate once dated your sister. The estimated 1.5 percent probability assumes an exhaustive search to find something in common with the person sitting next to you.

Six Degrees of Separation

The idea of coincidences has developed in the popular culture to a concept called "six degrees of separation," implying that everyone on Earth is related to everyone else through six degrees of separation. If you and the person next to you know one person in common, that is one degree of separation. If your daughter's best friend is married to the man next to you, that is two degrees of separation; and so on. An example of six degrees of separation might be something like the following:

One of your son's girlfriend's dentist's patients is married to a cousin of a friend of the woman next to you on the plane.

While, to my knowledge, no one has been able to prove definitively that the "six degrees of separation" hypothesis is true, it seems reasonable that it might be so. Why wouldn't everybody be related through such a drawn-out set of connections? But you might ask: "How might I ever have a connection with a farmer living in a remote province in China?" That's a very good question, and no one can really answer it. But I believe connections among countries work through well-connected individuals, who exist in all societies. The *New Yorker* writer Malcolm Gladwell discusses this phenomenon at length in his excellent book *The Tipping Point.* Certain people are natural born "connectors;" they seem to know everyone and act almost like airline hubs, connecting people from disparate geographic areas and walks of life.

Connections Among People of the World

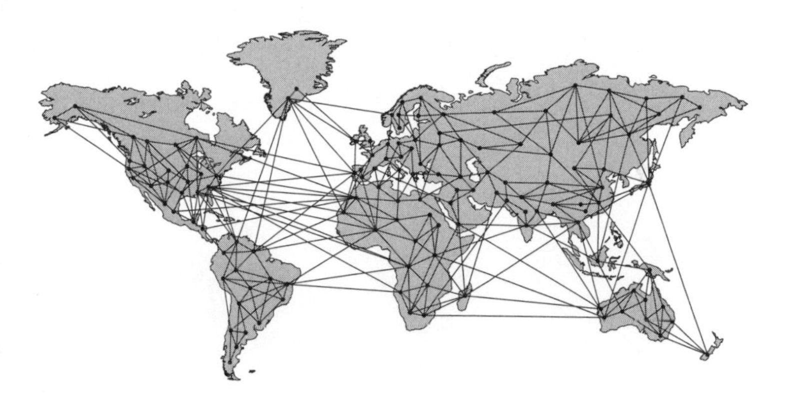

You can expend two or three degrees of separation from the farmer in China to an official connected with foreign trade to your uncle's business partner in California, whose assistant knows the Chinese official. But further research would have to be done to ascertain whether six degrees would do or whether we need seven or more to connect everyone on earth.

CHAPTER THIRTEEN

How to Succeed in Love (Find the Best Apartment, or Adopt the Best Puppy)

CHANCE AND TIME are intimately linked in our universe. Every day, chance occurrences take place that present us with new opportunities, challenges, and perils. To make the best of them requires a knowledge of how to make decisions in sequence. Nowhere is this more true than in the realm of love. A single person will meet a large but limited number of available members of the opposite sex over a lifetime, and a key question is: how do know when you've met the right person for you, the person you would like to marry? The assumption here is that you can't go back; once you've severed a relationship with someone, he or she is gone. How will you know which person is best for you?

Rejoice, because probability theory can help you find the

best possible match. And moreover, it can be applied to areas other than love. An apartment or house does not stay on the market for long, and often you have to make a decision on buying or renting before you've seen all your choices; how do you know you've found the best one? Jobs, too, don't stay on the market very long, and once you've been offered a position and rejected it, you can't go back. And cute puppies don't stay forever in a pet shop, and you may have to make a quick decision on whether this one is the best for you. So how do we know when to stop looking and choose?

A mathematical theorem has been developed that gives us the best sampling and stopping rule for all these situations. It can be found and further explained in books on probability.[16] But the strategy is as follows:

> You will maximize your probability of finding the best spouse if you date about thirty-seven percent of the available candidates in your life, and then choose to stay with the next candidate who is better than all previous ones.

Isn't It Romantic?

This is, indeed, a very strange-sounding rule. But mathematicians have proved it works better than any other. The number thirty-seven percent is an approximation of the exact number $1/e$, where e is the base for natural logarithms, or 2.71828 . . . Of course, this rule can't guarantee success, but, as Churchill said of democracy, it's the worst strategy except for all the others, and it gives you a thirty-seven percent probability of making the best decision. Any other strategy—whether choosing earlier or later—significantly decreases your probability of success in finding the best candidate. Here is one example. Suppose over a lifetime you expect to meet one hundred available candidates. If you marry the first one, the chance that you have indeed found the best of all one hundred candidates is only 1/100. Likewise, if you wait to meet all one hundred candidates, you will have rejected the ninety-nine who came before, and the possibility that the last person you meet is also the best is again only 1/100. The best strategies allow you to sample for a while, in order to learn about the various candidates; and of all such strategies, the best has you sampling thirty-seven percent of the total and then choosing the first candidate thereafter who beats all the ones who came before. Of course, there is a chance you will never find one who is better than all thirty-seven percent you've already seen. But nothing in love is certain, and this way you have the highest possible odds of finding your number one match. So if you are a young woman who expects to meet one hundred attractive bachelors over her dating years, you should let the first thirty-seven of them go, and marry the first one you meet thereafter who is more attractive to you than all thirty-seven young men you have already dated. Now, don't you wish your mother would give you advice like that?

CHAPTER FOURTEEN

How to Make Decisions Under Uncertainty

In the previous chapter we addressed one example of making sequential decisions under uncertainty. But in order to be able to make such decisions in a wide variety of complex situations, we need to learn something about "probabilistic averages."

> The expected value of a random quantity is simply the sum of the products of its values and their probabilities. In other words, the expected value is a weighted average—the more likely or unlikely an outcome, the more or less you value it.

Let's look at an example to show you what this means. Suppose someone offers you an investment that has a thirty percent chance of earning you one thousand dollars, a twenty percent chance of earning you two thousand dollars, and a fifty percent chance of losing you four hundred dollars. How much is this investment worth? Asked a different way, what do you *expect* to make on this investment? The answer is: $0.3 \times 1{,}000 + 0.2 \times 2{,}000 + 0.5 \times (-400) = \$500$. This is not to mean that, on a one-time basis, you should do it—there is a fifty percent chance of losing four hundred dollars here. But long term, if you had the opportunity of making such an investment every day, it would be worthwhile, since, on the average, you make five hundred dollars every time.

Fair Games

A fair game is defined as a game in which the expected payoff is zero. It is an extension of the popular idea of a zero-sum-game. In a fair game of chance the probabilities of winning and losing are such that *on the average*, what you win and what I win, assuming we're playing against each other, is zero. The game does not favor one of us over the other. Casino games are never fair games—they always give the house the advantage, even if the advantage is small. Of course, we've seen earlier that even in a fair game, the gambler will eventually lose when playing against an "infinitely wealthy" adversary, such as a casino.

Let's look at a simple game, roulette. If I'm betting on a number, my chances of winning are 1/38, because there are thirty-six numbers plus the zero and double zero. But if I win, I will get only thirty-six times the amount I bet. So my expected value for a single play is (assuming a bet of one dollar):

$[(-1) \times (37/38)] + [36 \times (1/38)] = -0.026$, or a loss of two-point-six cents, on average. If you keep playing such games long enough, you will lose everything, since you are losing about two-point-six cents with every spin of the wheel.

How to Value an Investment

Investments are no different from games of chance. Some of them have positive expected value and some negative; others have zero expected value and are similar to a "fair game." In order to calculate the expected value, we need to "weight" the various outcomes by their probabilities. For example, we may have an investment opportunity for which the following outcomes are possible:

PROFIT OR LOSS ($)	PROBABILITY
-1,000	0.1
-500	0.2
0	0.1
1,000	0.2
2,000	0.3
3,000	0.1

Note that the sum of the probabilities must be 1.0, or one hundred percent, as indeed it is in this example. We want to compute the expected profit or loss of this investment.

It is equal to:

$$(-1{,}000) \times 0.1 + (-500) \times 0.2 + (0) \times 0.1 + (1000) \times 0.2 + (2000) \times 0.3 + (3000) \times 0.1 = \$900$$

Just to make clear, this nine hundred dollars is on top of your initial investment, so that you ultimately end up with one thousand, nine hundred dollars, not just nine hundred dollars. Suppose that the cost of this investment is one thousand dollars. Are you better off making this investment or putting your one thousand dollars in the bank, where you can make five percent a year, which is fifty dollars? There is no unique answer to this question. Putting the money in the bank (assuming it is FDIC insured) would give you fifty dollars with certainty, while the other investment is speculative. Long-term, however, meaning if you can make such an investment again and again, you are better off with the investment since—on average—you gain nine hundred dollars every time you invest.

CHAPTER FIFTEEN

Game Strategies

THE THEORY OF GAMES IS SERIOUS STUFF. It has been developed to model conflict situations ranging from a game of chess to war to market competition. In a classic game, you make a move and then your opponent makes a move, and so on, until a final outcome is reached.

Simple games such as tic-tac-toe are perfectly solvable. As anyone who has played tic-tac-toe knows quite well, with the slightest amount of strategy it's impossible to lose (but then, it's impossible to win, also). Other games are more complicated. In chess, for example, moving first still gives an advantage but by no means determines the outcome of the game. Conflict situations such as competition in the market place, diplomatic maneuvering, and war are all modeled as games.

Be it chess, bridge, or war, surprise is an immensely important element in a game. If your opponent knows what you are going to do next, then protective action can be taken beforehand, foiling your move. Winning strategies are often *mixed strategies*, in which a certain move is made with a given probability, and another with another probability. We won't get into details in our elementary treatment here, but a simple example will illustrate the idea. If an army is to attack the enemy, the time of the attack should be chosen randomly, so that the enemy is not prepared for it. In the simplest case, suppose a general wants to strike the enemy sometime within the next twelve hours. To assure surprise, the general may roll a die to choose randomly a pair of hours from noon to midnight, and then toss a coin to choose which of two consecutive hours to use.

CHAPTER SIXTEEN

Bayes's Theorem

In 1761, an Anglican minister named Thomas Bayes attempted to use the theory of probability to prove the existence of God. He fiddled with equations of *conditional probabilities*—probabilities of events that are conditional on other events taking place—and managed to *reverse* the orders of the condition. The result made him pause. Did he have a mathematical proof of the existence of God? The reversal of the probabilities disturbed him greatly, and he put his paper in a drawer, refusing to look at it again. Two years later, in 1763, he died. His paper was discovered in the drawer and was published, causing a great stir in the mathematical community—a stir whose ripples continue to this very day. An entire field of science, now called *Bayesian Statistics*, was born as a result of the Rev. Bayes's discovery.

While he may not have proved the existence of God, Thomas Bayes nonetheless discovered an immensely important formula, and one with far-reaching and completely unexpected implications about scientific inference and how people process information.

Bayes's Law:

$$P(\text{A given B}) = \frac{P(\text{B given A}) \times P(\text{A})}{P(\text{B given A}) \times P(\text{A}) + P(\text{B given not A}) \times P(\text{not A})}$$

Two-and-a-half centuries after Bayes discovered his law, the formula is still controversial. Mathematically, the statement is perfectly correct. But it may sometimes lead to misinterpretations of data, if one is not careful enough. Bayesian analysis, done correctly, will lead to better results; but there are pitfalls to avoid. Let's look at a typical case of the use of Bayes's theorem.

Think of an illness, such as AIDS, for which tests exist that aren't perfectly reliable. Suppose that one-tenth of one percent of the population has the disease, and suppose that the test has a ninety-five percent reliability (which is quite good), meaning that if the person tested is sick, there is a ninety-five percent probability that the test will indicate so. Conversely, if the person is not sick, there is a five percent probability that the test will erroneously determine illness. Here is how Bayes's law is used in the analysis. We have the conditional probability that the test will say a sick person is actually sick: ninety-five percent; we have the *prior* probability that a person, randomly chosen from the entire population, is sick: one-tenth of one percent. We use Bayes's rule to determine the *posterior* probability that a person whose test results say he is sick actually is sick. At first glance you'd say the probability is ninety-five percent, right? After all, the test is ninety-five percent accurate. Well, just read on. According to Bayes's law we have:

$$P(\text{Sick } \textit{given} \text{ Test says sick}) = \frac{P(\text{Test will say sick given sick}) \times P(\text{Sick})}{[P(\text{Test says sick given Sick}) \times P(\text{Sick})] + [P(\text{Test says sick given not Sick}) \times P(\text{not Sick})]}$$

$$\frac{(0.95 \times 0.01)}{(0.95 \times 0.001) + (0.05 \times 0.999) = 0.0187}$$

This is, at first glance, a very surprising result. We have a very reliable test: a test that has a ninety-five percent probability of giving the right answer. We administer this test and get a positive result—a result indicating that the person is indeed sick. And yet, the posterior probability (meaning, the probability that a person whose test says he is sick actually is sick) is less than two percent! We might ask why we would ever administer such a test.

But the answer is more subtle. What Bayes's rule does is to mix the prior probability with the results of the test. If you think about it in the right way it makes sense: since such a small percentage of the population actually has AIDS (in our example), a positive test result is much more likely the result of a testing error than the person's actually having the disease. But, note, such an analysis assumes the person being tested is a typical sample of the general population—which brings us to the pitfalls mentioned earlier. Problems with the interpretation of Bayes's rule arise if we administer this test not to a random person in our entire population, but rather to someone in the hospital—someone who is more likely to be sick. Here the prior probability of one-tenth of one percent no longer applies. This, in fact, is the difficulty with using Bayes's rule: one has to be extremely careful in applying the prior probabilities and making sure they are valid. Often, prior probabilities—e.g., one-tenth of one percent of the population has AIDS—are subjective, while the conditional ones—e.g., the test has a ninety-five percent accuracy—are more objective. To get good results, one must be vary cautious. If the test is carried out in a hospital and one knows the percentage of people who are sick with this illness in the hospital, the appropriate prior probability should be used—not one that characterizes the population as a whole.

Here is a cute example of the use of Bayes's rule. A market researcher needs to interview married couples for a particular study. The researcher arrives at a house with three apartments. From the names on the mailboxes downstairs, the interviewer knows that only one of the three apartments has a married couple living in it. The others have two men, and two women living in them, respectively. But when the researcher goes upstairs to the apartments, the doors are unmarked and it is impossible to tell which of the three apartments is which. Therefore, the prior probability of knocking on the right door (that of the married couple) is clearly 1/3. The researcher knocks on the door, and a woman answers. What is *now* the probability that the researcher has reached the married couple? From an extended form of Bayes's rule, we have:

$$\text{P(Married couple given a Woman answers)} = \frac{\text{P(Woman answers given a Married couple)} \times \text{P (Married couple)}}{\begin{array}{c}\text{[P(Woman answers given a married couple)} \times \\ \text{P(Married couple)] + [P(Woman answers, given two men live} \\ \text{there)} \times \text{P(two men live there)] + [P(Woman answers given} \\ \text{two women live there)} \times \text{P(two women live there)]}\end{array}}$$

Now let's make some reasonable assumptions:

1. If the researcher reached the apartment with the married couple, then there is a 50-50 chance that either person answers the door, meaning that given that the married

couple was reached, the chance that a woman will answer is fifty percent.

2. If the researcher reached the apartment with the two men, then there is a zero probability that a woman will answer (we assume the men are unpopular and have no female visitors).

3. If the researcher reached the apartment with the two women, then there is a one hundred percent probability that a woman will answer the door (again, we assume no gentleman callers).

4. The prior probabilities—meaning the probabilities prior to the event that a woman has answered—of reaching the married couple, the two men, or the two women, are all 1/3 each.

We now plug in all these probabilities to the formula above and get that the posterior probability that the researcher has reached the married couple, given that a woman has already answered the door, is:

$$\frac{(1/2 \times 1/3)}{(1/2 \times 1/3) + (0 \times 1/3) + (1 \times 1/3)} = 1/3$$

This result is surprising. Before the researcher knocks on the door, the chance of having reached the married couple is 1/3. The researcher then gains some information: he knocks on the door and notes that a woman answers. It would seem that this bit of information should change the probability of having reached the right apartment, but it does not in this case. The final, posterior probability of having reached the married couple is still 1/3. Why? Because of the structure of this problem. There are three women here: one married and the other two living together. The probability that the researcher has reached the married couple is identical to the probability that the woman who answered the door is indeed the right woman out of the three who live in this building—which is one in three. The reader might find it useful and entertaining to try various made-up versions of this problem. Assume that the building has one apartment with seven men and three women, or one with four women and one man, etc., and try various numbers of apartments. See what you get for the probability that the researcher reaches the right apartment with the married couple, assuming a woman has answered the door.

Another interesting example of Bayes's law—and one that caused a controversy a few years ago when it was published in a syndicated newspaper column by Marilyn Vos Savant (Marilyn was right, and scores of Ph.D.s in math and sciences who wrote to complain about her solution were wrong), is the following. We use a somewhat different but equivalent version of the problem than the one published in the papers. This one is called the prisoner's problem.

Three prisoners share a cell. The warden tells them that one of them will be executed in the morning. One prisoner asks the warden, "If one of us will be executed, then clearly one of the other two prisoners in this cell will *not* be executed, right?" "Sure," says the warden. "OK," says the prisoner, "so please just point to one of these two." But the warden won't do it. "No," he says, "If I do that, then your chance of dying tomorrow will go up from 1/3, which it is right now, to 1/2."

This seems like a logical argument. It stands to reason that the space of possibilities for this problem will go down from three equally likely points to two equally likely points. But actually it does not. To see this, we will use Bayes's rule. Let's call our prisoner "Joe," and the others Peter and Paul.

$$P(\text{Joe dies given Warden points to Peter}) =$$

$$\frac{P(\text{Warden points to Peter given Joe dies})P(\text{Joe dies})}{\begin{array}{c}P(\text{Warden points to Peter given Joe dies})P(\text{Joe dies}) + \\ P(\text{Warden points to Peter given Peter dies})P(\text{Peter dies}) + \\ P(\text{Warden points to Peter given Paul dies})P(\text{Paul dies})\end{array}} =$$

$$\frac{(1/2)(1/3)}{[(1/2)(1/3) + (0)(1/3) + (1)(1/3)]} = 1/3$$

The same as the previous example.

Thus, despite what we may think, Joe's probability of dying remains the same—one-third only—despite the fact that the warden points to one of the two other men. Of course, since Joe's odds of dying remain at 1/3, Paul's odds of dying (assuming the warden pointed at Peter) increase to 2/3. See if you can use Bayes's law to prove this.

The version of this problem in Marilyn's column was the "Monty Hall Problem," based on the television program "Let's Make a Deal," in which a prize lies behind one of three doors. The probabilities there are identical to the ones in this example. The hidden Bayesian logic behind the problem has stumped many. Bayes's theorem remains a complicated, powerful, and often misunderstood tool in applied probability theory.

CHAPTER SEVENTEEN

The Normal Curve

THE MOST IMPORTANT PROBABILITY DISTRIBUTION is called the *Normal*, or *Gaussian* distribution. It has an interesting history. The name Gaussian commemorates the famous German mathematician Carl Friedrich Gauss (1777–1855), to whose credit are many mathematical discoveries of immense importance of both a pure and an applied nature. Gauss indeed used the normal distribution, and was thought to have been its discoverer. But in 1908 an English statistician by the name of Karl Pearson uncovered historical papers proving that the normal distribution was actually discovered a century earlier by the English mathematician Abraham De Moivre (1667–1754).

De Moivre was a French Huguenot who escaped the religious wars against his coreligionists and moved to England. There, he made the acquaintance of both Newton and Halley (of comet fame), and was even elected to the Royal Society. But he was not well-off and had to find occasional work as a tutor in mathematics to wealthy individuals.

De Moivre discovered the "normal law of errors." He found out that when many random factors accumulate, they form a bell-shaped curve, with less common values tailing off on either end, and the more average values grouping in the middle. The normal probability distribution, or the bell curve, is shown below.

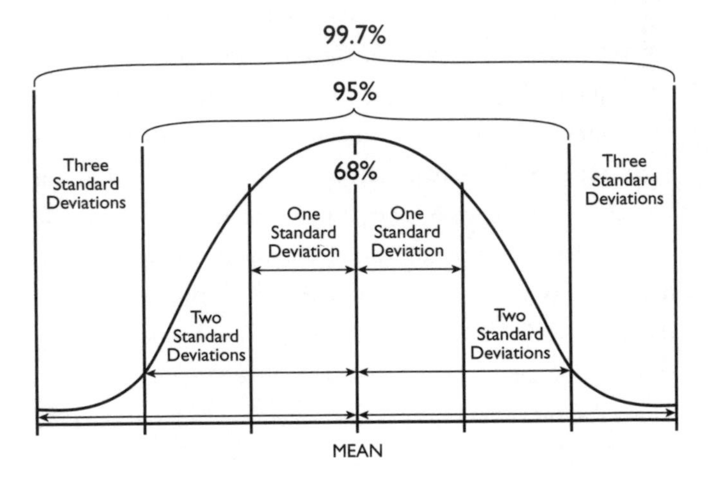

This curve has a *mean* (an average, or a center of mass if we assume a kind of "weight" attached to the area formed by the curve) and a *standard deviation*, which is a measure of the width of the curve (think of a standards deviation as a unit that measures horizontally across the curve). Since both the mean and the standard deviation can be any number we choose, there are infinitely many possible normal curves. Some of them are pictured below.

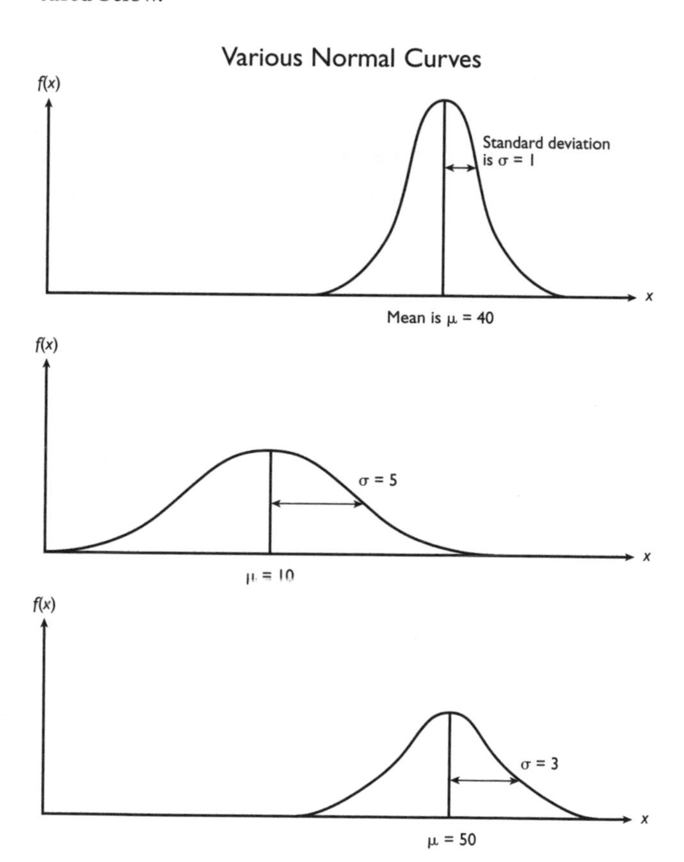

The normal curve seems to "appear out of nowhere" every time many random factors come together, and for this reason, ever since its discovery, people have attached almost mystical properties to it. In a way, the normal curve is a "divine law of nature," which tends to govern many phenomena in the real world. People's heights tend to be normally distributed around the average, for example, as are many other variables in nature. In 1960, Eugene Fama of the University of Chicago showed that percentage changes in the prices of stocks and other assets sold in efficient markets follow the normal distribution. This discovery led to an entire industry: quantitative financial analysis. Later, it was shown that exchange rates of currencies can also be modeled by a normal distribution.

The normal distribution is useful because its values and probabilities are tabulated, allowing us to carry out probability computations for any variable known to follow this distribution. In particular, here are some key values of this distribution.

Normal Distribution Probabilities

1. The probability that a normally distributed random quantity will fall within one standard deviation of the mean is about sixty-eight percent.

2. The probability that a normally distributed random quantity will fall within two standard deviations of the mean is about ninety-five percent.

3. The probability that a normally distributed random quantity will fall within three standard deviations of the mean is about 99.7 percent.

4. The probability that a normally distributed random quantity will fall within four standard deviations of the mean is close to certainty.

If the height of men in a certain population is normally distributed with an average of seventy inches and a standard deviation of four inches, then using the rules above we find the following:

1. The probability that a randomly chosen man has a height within sixty-six and seventy-four inches is about sixty eight percent.

2. The probability that a randomly chosen man has a height within sixty-two and seventy-eight inches is about ninety-five percent.

3. The probability that a randomly chosen man has a height within fifty-eight and eighty-two inches is about 99.7 percent.

The *tail* of the normal distribution is the area to the right of a point on the right side of the mean or to the left of a point on the left side of the mean. The name "tail" comes from an idea by W. Gossett, who drew the following amusing figure in 1908:

The Two Tails of the Normal Curve

Because the total probability is always 1.00, or one hundred percent, we can easily compute tail probabilities from the rules above.

1. The probability of falling to the right of the mean plus one standard deviation is about sixteen percent. (And, similarly, to the left of the mean minus one standard deviation.)

2. The probability of falling to the right of the mean plus two standard deviations is about 2.5 percent. (And, similarly, to the left of the mean minus one standard deviation.)

3. The probability of falling to the right of the mean plus three standard deviations is about 0.15 percent. (And, similarly, to the left of the mean minus one standard deviation.)

4. The probability of falling to the right of the mean plus four standard deviations or to the left of the mean minus four standard deviations is close to zero.

Here is an example. The daily exchange rate of the Euro to the U.S. Dollar over a certain period of time was normally distributed with a mean of one point two dollars per Euro and a standard deviation of zero point zero five dollars. What was the probability that the next day's exchange rate would be one point three dollars per Euro or more? From the rules above, we see that, since one point three is equal to two standard deviations to the right of the mean, the probability of falling at that point or higher—meaning the tail to the right of one point three—is 2.5 percent.

CHAPTER EIGHTEEN

Understanding the Presidential Election (and Other) Polls

THE MOST USEFUL FACT you should remember from the previous chapter is:

> For a normally distributed random quantity, the probability of falling within two standard deviations of the mean is about ninety-five percent.

(The precise value needed for *exactly* ninety-five percent probability is one point nine six standard deviations; so using two standard deviations gives us a slightly higher probability, whose exact value is 95.44 percent.)

This fact will serve us well in understanding polls and most other statistical sampling. When a poll is taken, the idea behind it is to try to obtain information about an entire population—such as the population of all voters in the United States—based on the information gleaned from a relatively small *random sample* from this population. By making every effort to assure that the sample is obtained at random, the researchers can then appeal to the laws of probability: because sampling outcomes from random polls follow the normal distribution.

The Mean and the Standard Deviation

Let's say you want to conduct a poll to measure the population's support for a political candidate. The sample size you're going to use is some number smaller than the population at large (obviously). If you imagine all the results the many, many different sample groups—i.e., potential polls—could yield, they form a normal distribution like the one below, with a majority clumped around the center, and a few extreme cases on either side.

The Normal Distribution of the Sample Proportion

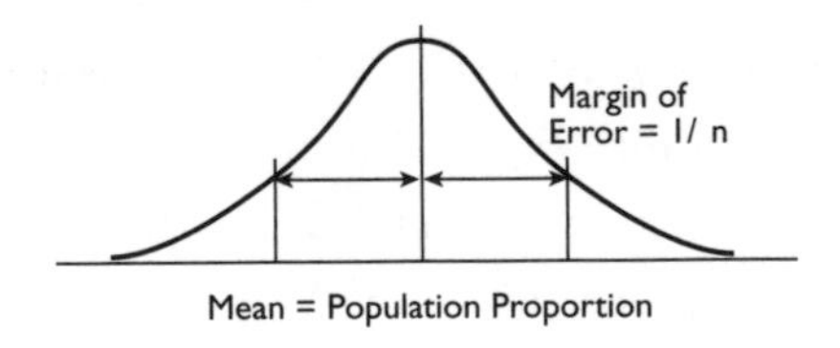

Now, we use the term *sample percentage* to mean the percentage of the sample group that chooses a particular answer (for example, if you ask ten people if they support a candidate, and six of them say yes, the sample percentage is sixty percent). We also use the term *sample proportion* to mean the sample percentage represented as a decimal (in other words, you divide the sample percentage by one hundred; so, in the above example, the sample portion is 0.6). Finally, we use the term *population percentage* to mean the actual percentage of the population at large that chooses a particular answer (so, if in reality sixty-three percent of the population supports a candidate, the population percentage is—you guessed it—sixty-three percent).

The mean of the normal distribution of sample percentages should equal the population percentage; the more samples you take, the greater the likelihood that you'll approach the population percentage (looking at the "bell curve" above we see how this makes sense). The standard deviation of the normal distribution of sample percentages is equal to:

$$\frac{\sqrt{(\text{sample proportion}) \times (\text{-1 sample proportion})}}{\text{sample size}}$$

This assumes the sample size is much smaller than the population size.

The numerator in the square root above is maximized when the sample proportion is equal to 0.5 (and one minus that quantity is also 0.5)—we maximize the term in order to cover ourselves in all cases, allowing the largest possible error to be considered. We define the *sampling error*, also called *margin of error*, of the survey as *two* times the standard deviation above. The number two is used to give us ninety-five percent probability of being correct in our poll results. Now, when we multiply:

$$\frac{2 \times \sqrt{0.5 \times (1 - 0.5)}}{\text{sample size}}$$

We get that it is equal to:

$$\frac{1}{\sqrt{\text{sample size}}}$$

This quantity, a priori—meaning before the sample is taken—maximizes our sampling error, and thus is used as the sampling error.

Sampling error (Margin of Error) for a poll with a ninety-five percent probability of being correct =

$$\frac{1}{\sqrt{\text{sample size}}}.$$

A poll result is reported as, for example: "Sixty-five percent for the candidate, with a sampling error of plus or minus five percent." In a typical report such as this one, we need to understand that (unless otherwise stated), the result is always true *with a ninety-five percent probability,* and the interpretation is that the actual percentage for the given candidate in the entire population from which the sample was drawn is *anywhere from the reported percentage to plus or minus the stated sampling error.* In this example, there is a ninety-five percent probability that the actual percentage of voters for the candidate in question is anywhere from sixty percent to seventy percent. Because of the sampling error, there is still a 2.5 percent probability that the actual percentage for the candidate is below sixty percent; and a 2.5 percent probability that it is above seventy percent.

You can also check whether what the pollsters tell you is true or not. (And, from the author's experience, pollsters often make mistakes in this respect.) Take the reported size of the random sample used in the poll; take the square root of that number and then compute one over this quantity. For example, suppose that the above poll was based on a random sample of one thousand people. In this case, the sampling error is:

$$\frac{1}{\sqrt{1{,}000}} = 0.0316$$

which is 3.16 percent, or roughly three percent, and not five percent as reported in the poll. In most serious polls, the sampling error will agree with the formula above.

CHAPTER NINETEEN

Conclusion

PROBABILITY THEORY is one of the most amazing inventions—or discoveries—of the human race. Probability is humanity's attempt to use pure mathematics to understand the un-understandable. It is our way of trying to learn something about the workings of chance. Chance remains forever untamable, for fate does what it wants to us and to the world around us. The universe is governed by laws that we may never quite fully understand. But the theory of probability gives us a window into the workings of randomness and of phenomena that, while we may not understand, with the aid of this little bit of mathematics, may be able to estimate in some way.

This book surveyed some of the basic elements of the theory of probability and—one hopes—has given you something to work with in trying to come to terms with the unpredictable, the unknowable, the untamable: nature's whim. Use it well.

A FEW PROBLEMS TO TEST YOUR NEW SKILLS (IN ORDER OF APPEARANCE OF TOPICS IN THE BOOK; ANSWERS FOLLOW BELOW)

1. What is the probability of drawing a club out of a well-shuffled deck?

2. What is the probability of drawing a face card (jack, queen, or king)?

3. What is the probability that the card drawn is either a face card or a spade?

4. What is the most probable outcome for the sum of the dots when two dice are thrown?

5. A card is drawn from a well-shuffled deck of fifty-two cards; the card is then returned to the deck, the deck is reshuffled, and a card is drawn again. What is the probability that both cards are spades? What is the answer if the first card is left out and not returned to the deck?

The following problem is from Frederick Mosteller's excellent book, *Fifty Challenging Problems in Probability*.[17]

6. My sock drawer has red and black socks only. When I reach into my drawer and randomly out pull two socks (without looking!), the chance that I get two red socks is exactly 1/2. What is the smallest number of socks in the drawer? Hint: Use the multiplication law to tell you what must be the probabilities that the first sock is red, and the conditional probability that the second sock is red given that the first one is red. Then look at the possibilities for number and color of socks in the drawer to figure out the smallest minimum number. (This is sampling without replacement.)

7. The *von Neumann Device:* Suppose that one of two people is to be randomly chosen to attend an important meeting. Someone claims that the coin to be used is not a fair coin, meaning that the probabilities of heads and tails are not fifty percent each (in fact, this person argues, no coin is fair—they're all biased in some way). How can I *still* use this coin to give me a device for creating two outcomes with equal probability of coming up, so I can choose fairly between the two candidates?

8. The *median* of a population is a number such that half the values in the population lie below it and half above it. If I ran-

domly sample two items from a population, what is the probability that the median of the population lies between them?

9. A bag contains eighty red balls and twenty black balls. A ball is drawn *twice, without replacement,* from the bag. What is the probability that both balls are red? Compare with the answer for sampling *with replacement.*

10. Try the de Finetti game on an event that interests you.

11. One in a hundred sugar cubes sold in boxes by the Holly Sugar Company is broken. If you reach for the sugar and randomly pick up two sugar cubes for your coffee, what is the probability that at least one of them is broken? (Assume independence. This assumption is inherent in all random sampling.)

12. According to the *New York Times* ("AT&T Wireless for Sale As Shakeout Starts," by Matt Richtel and Andrew Ross Sorkin, Wednesday, Jan. 21, 2004, p. C1), Verizon is the nation's largest cellular phone service provider, with thirty-six million subscribers out of a total of one hundred and forty-eight million cellular phone users in the United States. If six cell phone users are randomly selected on the street (recall that random selection implies independence), what is the probability that at least one of them is a Verizon subscriber?

Some Problems on the Probability of Dying from Various Kinds of Accidents

13. In 2003, there were 5,732 deaths from car accidents in France.[17] The population of France is 59,625,919. If I am going to live in France for five years, what is my probability of dying in a car crash?

14. According to *American Scientist* (data on *Americanscientist.org*), the probability of dying on a single one-leg flight, based on statistics for 1992 to 2001, including the fatalities that resulted from the September 11 terrorist attacks, is eight in one hundred million. If you are going to take twenty flights this year, what is the probability of surviving the year?

15. About a dozen people are killed in the United States each year as a result of being attacked by dogs.[18] Assume a population of two hundred and eighty million people over the next few years. What is the probability of being killed by a dog over twenty years?

16. For the United States, automobile fatality statistics are (for the most recent year of available data): 40,676 deaths from car crashes, out of a total population of two hundred and eighty million people. Compare the car fatality probability for one year in the U.S. and in France. What is the probability of dying from a car crash in the U.S. in the next twenty years?

17. If ten people are gathered in a room, what is the probability of at least one birthday match?

18. If nine hundred categories exist, what is the number of people needed for ninety-five percent probability of a match?

19. If you expect to date fifty people over the next five years and want to get married within this period of time, what is the best strategy for finding the best mate?

20. If one investment can give you one thousand, two hundred dollars with probability 0.2 and you could lose eight hundred dollars with probability 0.8, and another investment can give you six hundred dollars with probability 0.7 and a loss of three hundred dollars with probability 0.3, which investment is better?

21. There are five apartments. One has two men, one has a woman and two men, one has two women and three men, one has six women and one man, and one has a married couple. If I knock on the door and a woman answers, what is the probability that I've reached the one with the married couple?

22. In a poll, two thousand people are asked their opinion on a yes-no question. What is the margin of error?

ANSWERS

1. $13/52 = 1/4$ = twenty-five percent.

2. $12/52 = 3/13 \approx$ twenty-three percent.

3. $12/52 + 13/52 - 3/52 = 22/52$.

4. The most probable sum is seven, with probability $6/36=1/6$.

5. $1/4 \times 1/4 = 1/16 = 0.0625$. Without replacement: $1/4 \times 12/52 \approx 0.0588$.

6. We must have $p_1 p_2 = 1/2$. The possibility with the smallest number of socks is: RRRB, since then $p_1 = 3/4$ and $p_2 = 2/3$, so that $p_1 p_2 = 2/4 = 1/2$.

7. Call the probability of heads p and of tails $(1-p)$; toss the coin twice. Ignore HH and TT; use HT for one choice and TH for the other: both have equal probability of $p(1-p) = p-p^2$.

8. There are four possibilities: either both values fell below the median (BB), both above the median (AA), or one above and the other below (two cases: AB, BA). Since the chance of falling on either side of the median is by definition 1/2, each one of the pairs has probability $(1/2) \times (1/2) = 1/4$ (by independence). Two of these possibilities, AB and BA, capture the median between them. Hence the probability is $\times + \times = 1/2$.

9. $80/100 \times 79/99 = 0.63838$. With replacement: $0.8 \times 0.8 = 0.64$, which is slightly higher.

11. $1 - (0.99 \times 0.99) = 0.0199$.

12. $1 - (112/148)^6 = 0.8121$.

13. $1 - (0.9999)^5 = 0.0005$, or one-half of one percent.

14. $1 - (99{,}999{,}992/100{,}000{,}000)^{20} = 0.9999984$. Not bad!

15. $1 - (279{,}999{,}988/280{,}000{,}000)^{20} = 0.00000086$.

16. The probabilities for France and the U.S. are about the same, one in ten thousand per year. The probability of dying in a car crash over the next twenty years is:

$$1 - \frac{(280{,}000{,}000 - 40{,}676)^{20}}{280{,}000{,}000}$$

= 0.003, or one-third of a percent. This is not an insignificantly small probability any more, once again showing how probabilities compound as the number of trials (here, years of living) grows.

17. 1 − [(364/365) x (363/365) x (362/365) x (361/365) x (360/365) x (359/365) x (358/365) x (357/365) x (356/365)] = about twelve percent.

18. $\sqrt{900} \times 1.6 = 48$.

19. 0.37 x 50 = 18. After dating eighteen people, propose to the first person better than the initial eighteen.

20. Expected return of the first investment = ($1,200 x 0.2 + (-$800 x 0.8) = -$400. Expected return of the second investment = ($600 x 0.7) + (-$300 x 0.3) = $330.

21. This problem is deceptively simple. Since there are ten women who live in the apartment complex, and one of them is married, there is a 1/10 chance the woman you meet is the married one. So, the answer is 1/10.

22.

$$\frac{1}{\sqrt{1{,}000}} = 2.23\%.$$

NOTES

1. See F. N. David, *Games, Gods and Gambling*, New York: Dover, 1998, p. 7.
2. See Nachum L. Rabinovitch, "Probability in the Talmud," *Biometrika* 56, no. 2 (1969), pp. 437–41.
3. Sir Thomas Heath, *A History of Greek Mathematics*, Vol. I, New York: Dover, 1981.
4. Amir D. Aczel, *Statistics: Concepts and Applications*, Chicago: Irwin, 1995, p. 72.
5. What is a well-shuffled deck? How many times should one shuffle it? The statistician Persi Diaconis of Harvard proved some years ago that seven times is the minimum required number of shuffles to give a reasonable assurance of randomness. Of course, the shuffle should be done well. See the appendix for more on well-shuffled decks.
6. See the book by Edward Thorpe, *Beat the Dealer*, 1961, for a discussion of counting cards in blackjack.
7. David Freedman, Robert Pisani, Roger Purves, and Ani Adhikari, *Statistics*, 2d ed., New York: W.W. Norton, 1991, p. 234.
8. N. H. Josephy and Amir D. Aczel, "A Note on a Journal Selection

Problem," *Methods and Models of Operations Research*, 34 (1990), pp. 469–76.

9. As attested by the title and contents of Burton Malkiel's famous book, *A Random Walk Down Wall Street*, New York: W.W. Norton, 1995.
10. William Feller, *An Introduction to Probability Theory and Its Applications*, Vol. 1, New York: Wiley, 1973, p. 87.
11. Lester E. Dubins and Leonard J. Savage, *How to Gamble if You Must*, 1965, reprinted as *Inequalities for Stochastic Processes*, New York: Dover, 1976.
12. This example is from William Feller, *An Introduction to Probability Theory and Its Applications*, Vol. 1, New York: Wiley, 1973.
13. This test is described, for example, in Amir D. Aczel, *Complete Business Statistics*, 5th ed., New York: McGraw-Hill, 2000, p. 636.
14. GE "FloodLight R30."
15. This information is based on a lecture delivered at the mathematics department at the University of California at Berkeley on August 10, 1998, by the Harvard (then Stanford) mathematician Persi Diaconis.
16. See Frederick Mosteller, *Fifty Challenging Problems in Probability*, New York: Dover, 1987, p. 74.
17. Ibid., p.1.
18. Elaine Sciolino, "Garçon! The Check, Please, and Wrap Up the Bordelais!" *The New York Times*, January 26, 2004, p. A4.
19. James Gorman, "Life-and-Death Decisions About Four-Legged Prisoners," *The New York Times*, January 26, 2004, p. B5.

APPENDIX

by Brad Johnson

Shuffling

A fresh pack of playing cards is arranged in sequence by suit, with the jokers and spare cards at one end. After one breaks the seal and removes the extra cards, the deck is shuffled to randomize the sequence of the cards. Let's define a *random shuffle* as one which is equally likely to leave the deck in any one of the possible configurations (permutations) of the fifty-two cards: a *well-shuffled deck.*

The average card player's shuffling routine doesn't always approximate a random shuffle. When computer-randomized decks were introduced to tournament bridge play in the 1970s, there was an outcry from players objecting to a rise in strangely distributed hands. But it turned out that it was the players who'd been at fault. They'd been insufficiently shuffling the decks, resulting in a greater number of evenly distributed hands than would be expected from well-shuffled decks.

The two common types of shuffling which players use are the riffle or faro shuffle, in which the deck is split in half and the cards are riffled together; and the exchange or overhand shuffle, in which chunks of cards are pulled out and slid back in randomly.

Mathematicians have studied the problem and determined that the overhand shuffle is a terrible randomizer, requiring hundreds of iterations for randomness. Six or seven riffle shuffles are sufficient to produce a well-shuffled deck. What's remarkable is that fewer than five shuffles is notably insufficient. The transition from order to randomness is rapid, much as clumps of flour disappear almost magically into batter with just enough turns of the spoon.

It is straightforward to show that with only five riffle shuffles, all the configurations of a fifty-two card deck are not equally likely. In fact, some configurations cannot be reached! We can see this by thinking about what kinds of sequences a riffle shuffle produces.

Consider an eight-card deck which is in order: 12345678. After one riffle shuffle the following sequence is possible: 12563784. This sequence consists of two *rising sequences*, the longest possible sequences of cards in increasing order (not necessarily touching). The sequences are 1234 and 5678.

Starting with a sorted deck, a single riffle shuffle can only produce two rising sequences. The next riffle shuffle can produce at most four, by splitting the two rising sequences created by the first shuffle. Each successive shuffle can at most double the number of rising sequences. Therefore the third shuffle can produce eight, the fourth sixteen, the fifth thirty-two rising sequences. Because a fifty-two card deck in reverse order consists of fifty-two one-card rising sequences (and fifty-two is greater than thirty-two), it is impossible to produce a reversed deck with five riffle shuffles.

Using different measures of sufficient randomness, mathematicians have argued that either six or seven shuffles is the magic number after which further shuffles produce no discernable randomization.

Persi Diaconis, in 1986 (with David Aldous, refined with Dave Bayer in 1992), measured the expected gain of a perfect gambler who could take advantage of residual patterns left in the deck from inadequate shuffling. The gain plummets to near zero by seven shuffles. In 2000, Nick and Lloyd Trefethen took a different approach. They used the information theory of Claude Shannon (who once tried to beat roulette and later made millions on the stock market). If you have a deck in which you know the order of the cards, you have total information. With a perfectly shuffled deck, you have zero information. Imagine trying to send a message using a coding system based on what you know about the deck; information theory states that as long as you have some information, you can send the message given enough time. Using computer simulations, the Trefethens found that the amount of information in a deck after six riffle shuffles is practically zero.

What does this mean when you sit down to play cards?

Learn how to riffle shuffle. Don't use the overhand shuffle if you plan to play based on the assumption of a well-shuffled deck. And shuffle the deck at least a half-dozen times. Anyone who shuffles the deck more is holding up the game.

Expected gain

The essential mathematical concept behind gambling—and the stock market, and all other fields of risk and reward—is that of *expected gain*. Expected gain ties the *probability* of outcomes (the risk) to the *desirability* of outcomes (the reward). The expected gain of a decision A, given some cost C for that choice and some reward R for a positive outcome, is:

$$EG(A) = P(A) \times R(A) - C(A)$$

The expected gain of, say, choosing to marry a particular person, is quite difficult to determine and is rather subjective; how do you measure the costs and rewards of such a decision? Gambling eliminates much of the complexity by giving exact dollar amounts to the cost and reward. For example, if you have a thousand dollars to invest with a twenty percent probability of doubling your money (or losing it all), your expected gain is:

$$EG = .20 \times \$2{,}000 - \$1{,}000 = -\$600$$

Conversely, if the probability of success of doubling your money is eighty percent, your expected gain is:

$$E\,G = .80 \times \$2{,}000 - \$1{,}000 = \$600$$

Clearly the probability of success in the second case makes the investment worthwhile, while the probability in the second doesn't.

The world of the casinos includes a multitude of gambling opportunities, from flashing slot machines to elite baccarat tables and off-track betting. Although each game has its own distinct set of rules and rituals, they all have one thing in common: the house always wins. More specifically, all casino games are designed so that, on average, players lose more money than they win. They are paying to play.

Casino games can be separated into two distinct classes: unwinnable and winnable games. Most casino games, including the slots, keno, craps, pai gow poker, baccarat, and roulette, are mathematically impossible to beat. Different playing and betting strategies will affect the rate at which a player can expect to lose money, but all players will expect lose with time. A few casino games, including blackjack and some variants of video poker, have a positive expected gain under optimal play (though casinos bar blackjack players who use optimal strategy—see, for example, the section on counting cards in blackjack on pg. 16). The games in which the player is effectively competing against other players, and the casino takes a guaranteed cut—namely poker and horse and sports betting—offer a genuine possibility for skilled players to profit.

In a fair game, the given odds reflect the true odds.

A *fair game* is one in which the expected gain of the player is zero. On average, the player will win as often as he loses. Remember, however, the Gambler's Ruin Theorem (pg. 45). By the formula for expected gain above, a game is fair when

$$\frac{C(A)}{R(A)} = P(A)$$

That is, when the cost-to-reward ratio is the same as the probability of success. Gamblers usually use the terminology of *odds* instead of probability, comparing the *given odds* to the *true odds.* If the given odds are the same as the true odds, the game is fair. Not surprisingly, in all casino games where the player competes against the house, the given odds do not reflect the true odds, and the player's expected gain is negative.

Having to juggle given odds, true odds, and probabilities certainly makes probabilistic reasoning more difficult. This is yet another trick that works against the gambler, who may simply throw up his hands at the arithmetical effort and place a bet on Sky Jumper because he likes the name of the horse.

But if you can reduce all these odds to obtain a value for the expected gain, you will know whether the game is fair (EG = 0), loaded in your favor (positive EG), or stacked against you (negative EG).

Odds

Betting games don't use the terminology of probability and expected gain; they use the terminology of odds and payoffs. "Red/Black pays 1-to-1 (even money)." "Sky Jumper is the 3-to-5 favorite." Here, "1-to-1" and "3-to-5" represent the given odds of the bet, and are in the format of *payoff*-to-*cost* of a bet.

With given odds of 3-to-5, a winning five dollar bet returns eight dollars: the initial bet of five dollars plus a payoff of three dollars. A winning even-money bet delivers a payoff of one dollar for every dollar bet.

What is the relationship between the odds and probability? We see that for odds of *payoff-A*-to-*cost-B* (*A* for *ante*, *B* for *bet*) the reward R = cost B + payoff A. Rewriting the expected gain formula in terms of A,

$$EG = P(B + A) - B$$

For a fair game, where the expected gain is zero (and the true odds equal the given odds),

$$P = \frac{B}{(B + A)}$$

Which can be rewritten in terms of the given odds and true odds as

$$\frac{A}{B} = \frac{(1 - P)}{P}$$

Note that $1 - P(X)$ is another way of writing the probability of not getting X. So the definition for the true odds in terms of the probability is

$$\text{True odds of X} = \frac{\text{P(not X)}}{\text{P(X)}}$$

So if the given odds for our horse Sky Jumper are fair (a point we will discuss later), Sky Jumper would be expected to win such a race five times for every three times he loses, a probability of success of

$$\frac{5}{(5+3)} = 5/8$$

If someone offers you an even-money bet (1-to-1 odds), ask yourself, "Am I as likely to win as I am to lose?" If the answer is yes, it is a fair bet, with zero expected gain for both players. The probability of success for each player is

$$\frac{1}{(1 + 1)} = 1/2$$

Roulette

Roulette is the best friend of the author attempting to discuss the mathematics of gambling. It is gambling stripped to its mathematical essentials. Most games in the casino have fancy rules or sequences of play, but roulette is just one big wheel.

In American roulette, the wheel and betting board look like this:

There are eighteen red spaces, eighteen black spaces, and the two green zero and double zero spaces. You place a bet that the ball will land on a particular number or range of numbers. If your bet succeeds, you win at the odds shown. Otherwise, you win nothing.

If you place one dollar on red to win, and the ball lands on red, you receive two dollars. The given odds are 1-to-1, also known as even money.

Your cost-to-reward ratio is 1/2. If roulette were a fair game, the likelihood of the ball landing on red would be 1/2. But as there are eighteen red spaces and twenty non-red spaces, the true odds are 20-to-18; the probability of success is 18/38, which is less than 1/2. The expected gain is minus 1/19 (minus 0.0526). For an average bet of one dollar on red or black, you can expect to lose a little more than a nickle. The table on the following page shows the probability of success, the true odds, given odds, and expected gain for all of the possible roulette bets:

Bet	*Name*	*Probability*	*True Odds*	*Given Odds*	*EG*
1 number	Straight Up	1/38	37 to 1	35 to 1	-0.0526
2 numbers	Split	2/38 (1/19) 1	8 to 1	17 to 1	-0.0526
3 numbers	3 Line	3/38	35 to 3	11 to 1 (33 to 3)	-0.0526
4 numbers	Corner	4/38 (2/19)	17 to 2	8 to 1 (16 to 2)	-0.0526
5 numbers	First Five	5/38	33 to 5	6 to 1 (30 to 5)	-0.0789
6 numbers	First Six	6/38 (3/19)	16 to 3	5 to 1 (15 to 3)	-0.0526
12 numbers	Column or Dozen	12/38 (6/19)	13 to 6	2 to 1 (12 to 6)	-0.0526
18 numbers	Even/Odd, Red/Black, Low/High	18/38 (9/19)	10 to 9	1 to 1 (9 to 9)	-0.0526

What does this chart tell you? No matter the bet, your expected gain is negative. You can't beat roulette.

How to Beat Roulette

I admit it: it actually is possible to beat roulette. People have done it. How? Not by beating its mathematics, but by beating its physics. Edward Thorp, the mathematician best known for his book about counting cards to win at blackjack, *Beat the Dealer*, also developed Newtonian formulas to describe the motion of the roulette wheel and ball. His calculations showed that by timing the spinning wheel and bouncing ball over successive spins and feeding the information into a computer, one could predict where the ball would land with sufficient accuracy to win at roulette. In 1962, Thorp teamed up with Claude Shannon, the inventor of information theory, to build a cigarette-box-sized analog computer and attack the Riviera Hotel on the Vegas strip (it was a great time to be a mathematician). Their jerry-rigged contraption suffered mechanical glitches, and Shannon and Thorp put the roulette-beating game aside.

In 1980, a group of West Coast physics nerds joined together as Eudaemonic Enterprises and developed shoe-heel digital computers which allowed them to successfully beat roulette. Their story is chronicled in Thomas Bass's *Eudaemonic Pie*. A postscript: Thorp and the Eudaemonic kids independently realized that the stock market was a much better game to play (more money, fewer kneecaps broken), and became multi-millionaires using computerized trading systems. These stories are told in *Beat the Market* and *The Predictors*.

Poker

Poker is the most popular American card game, first in many editions of the venerable *According to Hoyle*. With roots in the ancient Persian game, *As Nâs*, as well as card games of the Renaissance, the game poque came to the wild port of New Orleans in the nineteenth century. During the Civil War, draw and stud poker were invented. In ensuing decades further variations were invented and its place in American culture was cemented. The 1970s and 1980s saw the publication of the first serious poker books by professional players, such as Doyle Brunson's *Super System* and David Sklansky's *Theory of Poker*. In the 1990s the explosion of casinos and online play greatly expanded the ranks of professional and high-level amateur players.

There are many variations of the game poker, but the goal is always to have the best five-card poker hand—or to bluff everyone else in to folding. The hands are ranked in inverse proportion to the likelihood of achieving them (assuming no wild cards). Poker hands come in three categories. The first is cards of the same rank: one pair, two pair, three-of-a-kind, four-of-a-kind, and full house (a pair and a three-of-a-kind). The second is five cards of the same suit, known as a *flush*. The third is five cards in sequence, known as a *straight* (the ace can be high or low in straights, but not both, so A, 2, 3, 4, 5 and 10, J, Q, K, A are valid straights, and Q, K, A, 2, 3 is not). Five cards of the same sequence in the same suit is a straight flush, naturally. The *royal flush* (10-J-Q-K-A in one suit) is the highest possible hand. The odds are 649,739 to 1 against being dealt a royal flush.

Poker-Hand Probabilities

HAND	NUMBER	PROBABILITY	ODDS
One pair	1,098,240	0.4226	1.366:1
Two pairs	123,552	0.0475	20:1
Three of a kind	54,912	0.0211	46:1
Straight	10,200	0.0039	254:1
Flush	5,108	0.0020	508:1
Full house	3,744	0.0014	693:1
Four of a kind	624	0.00024	4,164:1
Straight flush	36	0.000014	72,192:1
Royal flush	4	0.0000015	649,739:1

The inclusion of wild cards, such as one joker (the *bug*) or two, introduce the possibility of a five-of-a-kind, which beats all other hands, and changes the probabilities of the hands in a complicated manner.

Every variation of poker has a *pot* of community money which is built through rounds of betting and then taken by the player with the winning hand. In each round of betting, players have the choice of placing a bet which opponents must match to stay in (betting nothing is known as *checking*); *calling*, or, matching the bet already made; matching the bet and *raising* by adding an additional bet; or *dropping* out of the betting (known as *folding*, in stud poker). The betting interval ends once all active players have equalized their bets.

There are numerous popular variants of poker, including five-card draw, seven-card stud, lowball, and Texas hold'em. The latter involves two cards dealt face down to each player and five community cards dealt in rounds of three cards, one card, and one card, for a total of four rounds of betting. This variation has become wildly popular, spurred by the ever-increasing popularity of Binion's World Series of Poker, an open poker tournament of Texas hold'em begun in 1970 and now with a multimillion-dollar jackpot.

The following is a brief introduction to a few of the essential mathematical concepts that underlie all variations of poker: pot odds, the Fundamental Theorem of Poker, and bluffing.

Pot odds

In poker, your expected gain for a given bet B and a given pot size A is:

$$EG = P(\text{of winning}) \times (A + B) - B$$

The *true odds* are:

$$\frac{(1-P)}{P}$$

The *given odds*, also known as the *pot odds*, are:

$$\frac{A}{B}$$

You want to bet so that your expected gain is positive, which is when the true odds are no more than the pot odds.

So if you think your odds of success are 4-to-1, you want the pot odds to be better than 4-to-1. If the pot is one hundred dollars, you want to bet less than twenty-five dollars. Conversely, you want the pot odds to be worse for your opponents than his true odds of success. Say you hold a full house and your opponent has a four-flush in five-card draw. If he draws one card, he has a nineteen percent chance of making the flush and beating your full house. He has 4-to-1 odds of making his flush (note that since his probability of success is about 1/5—i.e., he'll succeed one time for every four times he fails—the odds are 4-to-1). Say the pot is one hundred dollars. If you bet fifty dollars, he has to bet fifty dollars for the chance at a one hundred and fifty dollar pot: 3-to-1 pot odds. With your bet, you've made your opponent's pot odds worse than his given odds.

Calculating your given odds of success is complicated by the fact that you don't know exactly what your opponents hold. Calculating pot odds is complicated by multiple rounds of betting and deals in games like seven-card stud and Texas hold'em. When betting, you should consider not only the size of the present pot but how the pot may grow.

Fundamental Theorem of Poker

The Fundamental Theorem of Poker, according to David Sklansky's *The Theory of Poker* (Two Plus Two Publishing, 2002) is:

> Every time you play a hand differently from the way you would have played it if you could see all your opponents' cards, they gain; and every time you play your hand the same way you would have played it if you could see all their cards, they lose. Conversely, every time opponents play their hands differently from the way they would have if they could see all your cards, you gain; and every time they play their hands the same way they would have played if they could see all your cards, you lose.

The *gain* referred to in the theorem is *expected gain*. The theorem refers to the fact that when you miscalculate the true odds, your expected gain decreases; when your opponents miscalculate the true odds, your expected gain increases.

In other words, what makes poker interesting is that it is a game of incomplete information. You don't know which cards you will draw or which will turn up in stud, nor do you know your opponents' cards. The goal of poker (and any game where multiple players compete given imperfect knowledge) is twofold: to play as close as possible to how you would play given complete information, while tricking your opponents into playing based on faulty assumptions—which you do through bluffing.

A simple example of the Fundamental Theorem at work is in five-card draw. If you start with a pair in your hand and you draw three cards, the probability of improving your hand (either by making three-of-a-kind, four-of-a-kind, two pair, or a full house) is 0.287. If you draw two cards, the probability of improving your hand is 0.260. Since drawing three cards gives you a better chance of improvement than two cards, why would you ever draw only two cards?

You do that to bluff your opponent into thinking that instead of a pair, you hold three-of-a-kind before the draw. Say he also has a pair or two pair after the draw; if he thinks you have three-of-a-kind, he'll likely fold. He might even fold if he has a low three-of-a-kind. According to the Fundamental Theorem of Poker, you gain, since your opponent's actions would be different if he knew your cards.

Furthermore, if your opponent knows that you sometimes bluff by drawing two cards to a pair, when you do get a three-of-a-kind and draw two, he may think you only have a pair and call you with only a pair or two pair in his hand. You again would gain by the Fundamental Theorem of Poker.

Bluffing

So how often should you bluff? That depends on how wily your opponent is; if he believes every bluff you make, then bluff with impunity. If he calls every bluff, then don't bluff (play "tight") but bet heavily whenever you get a strong hand. Say, however, he's as clever a player as you. In that case, the likelihood of a bluff should reflect the pot odds. If your opponent's pot odds are 5-to-1 (say he has to bet twenty dollars into a one hundred dollar pot to call you), then the odds against your bluff should be 5-to-1.

The probability of getting a pair in the initial deal in five-card draw is 0.4226; the probability of getting three-of-a-kind is 0.0211. So for every twenty pairs you are dealt, you are dealt one three-of-a-kind. Your comparative odds are 20-to-1. If you always drew two cards to a pair or three-of-a-kind, your opponent would know that it was a 20-to-1 shot you weren't bluffing. But if you drew two cards to a pair half of the time, it would only be a 10-to-1 shot.

Drawing two cards to a pair one out of forty times, you're only bluffing once for every two times you are dealt a three-of-a-kind. In that case, the likelihood that you would have at least two pair after the draw is seventy-five percent (it's seventy percent that you have three-of-a-kind or better). It's 3-to-1 odds that you have a pair, and 7-to-3 odds that you have one or two pair. If the pot odds are worse than that, your opponent shouldn't call your bluff.

Say you only draw two cards to a pair if you are also holding the ace of spades and the three of clubs, a situation that arises once out of every hundred pairs you are dealt. Then the odds of your having a three-of-a-kind or better after the draw is just about 5-to-1.

Horse racing

> Then all together, they raised their whips above their horses, lashed them with the reins, and shouted words of encouragement to urge them forward. The horses raced off quickly, galloping swiftly from the ships. Under their chests dust came up, hanging there like storm clouds in a whirlwind. In the rushing air their manes streamed back. The chariots, at one moment, would skim across the nourishing earth, then, in another, would bounce high in the air. Their drivers stood up in the chariots, hearts pounding, as they strove for victory. Each man shouted out, calling his horses, as they flew along that dusty plain.
>
> —Book xxiii, *Iliad*

The roots of horse racing go deep. A Hittite document from 1500 BC details horse breeding and training. The Greeks held chariot races in the 23rd Olympiad, and began races with mounted horses in the 33rd Olympiad. The Boeotians even named one of their months *Hippodromius*—the "month of horse racing."

Horse racing deserves its sobriquet as the Sport of Kings—

the races of today descend directly from those set up by King James I in seventeenth century England and formalized by King Charles II a century later. In 1779 and 1780, the first races with dedicated prize money, the Epsom Oaks and Derby races, were established by the 12th earl of Derby. All aspects of the business can be ruinously expensive—from the gambler to the horse owner, who spends hundreds of thousands of dollars to buy and train one thoroughbred horse. All thoroughbreds are descended from one of three Near East stallions: the Byerly Turk, the Darley Arabian, and the Godolphian Barb, which were bred in England at the beginning of the 1700s.

Horse racing in the United States involves six to twelve horses racing counterclockwise around an oval track. The races are separated by breed: thoroughbred, standard bred (harness), and quarterhorse. Races are further categorized by distance (6 furlongs or 3/4 of a mile, 7 furlongs or 7/8 of a mile, one mile, 1 1/8 mile, 1 1/4 mile, 1 1/2 mile), stakes (the Belmont Stakes, for example, has a purse of one million dollars), sex (female, male, gelding), age (all racing horses have the same official birthday of January 1—as decreed by the Thoroughbred Racing Association), and horse quality. The races are Stakes, Handicap, Claiming, Optional Claiming, Allowance, Classified Allowance, and Maiden. Stakes races are the premier races: open races for which horses must qualify and compete for an established purse (or stakes). In Handicap races the horses are handicapped by having to carry weights. In Claiming races the horses are up for sale; they can be "claimed" at a given price. In Allowance races, there is some particular allowance for which horses are eligible, such as "2-years that have won no more than one race." Maiden races are for horses that have never won before.

The standard horsetrack bets, each of which delivers distinct payoffs, are Win, Place (first or second), Show (first, second, or third), and Across the Board (shorthand for combined Win, Place, and Show bets). The "exotic" bets involve selecting correct combinations of winners: the Quinella (the first two horses), Exacta (first two in exact order), Trifecta (first three in exact order), Superfecta (first four in exact order), the Daily Double (winner of two consecutive races), Pick Three (three consecutive races), and further variants.

The pari-mutuel system

Horse racing bets use the pari-mutuel system. The system was invented by Parisian perfume maker Pierre Oller in 1865, asked by a bookmaker friend to devise a system that guarantees a profit for him while remaining fair to the bettors. Oller realized that a percentage could be taken out from a pool of all bets, then payoffs made from the remaining pool, if the payoff odds are inversely proportional to the amount bet on each horse. In the pari-mutuel system bettors are competing against each other, not against the house.

As bets are placed on horses, the odds are adjusted (payoffs go down for favorites, up for longshots), less the "takeout" (usually at least fifteen percent), and also less a rounding down to standard payoff odds, known as the "breakage." The breakage has an average cost to the bettor of two percent.

Effect of breakage on payoffs on $2.00 bets

ODDS	PAYOFF RANGE	MAXIMUM EFFECT OF BREAKAGE	AVERAGE EFFECT OF BREAKAGE
1/20	$2.10 (minimum)	none	none
1/10	$2.10	-8.6%	-4.3%
1/5	$2.40-$2.60	-7.3%	-3.5%
2/5	$2.80	-6.4%	-3.2%
1/2	$3.00	-6.0%	-3.0%
3/5	$3.20-$3.40	-5.6%	-2.7%
4/5	$3.60-$3.80	-5.0%	-2.4%
1	$4.00-$4.20	-4.5%	-2.2%
6/5	$4.40-$4.60	-4.1%	-2.1%
7/5	$4.80	-3.8%	-1.9%
3/2	$5.00	-3.7%	-1.8%
8/5	$5.20-$5.40	-3.5%	-1.7%
9/5	$5.60-$5.80	-3.3%	-1.6%
2	$6.00-$6.80	-3.1%	-1.45%
5/2	$7.00-$7.80	-2.6%	-1.26%
3	$8.00-$8.80	-2.3%	-1.11%
7/2	$9.00-$9.80	-2.1%	-0.99%
4	$10.00-$10.80	-1.9%	-0.90%

ODDS	PAYOFF RANGE	MAXIMUM EFFECT OF BREAKAGE	AVERAGE EFFECT OF BREAKAGE
9/2	$11.00-$11.80	-1.7%	-0.82%
5	$12.00-$13.80	-1.5%	-0.73%
6	$14.00-$15.80	-1.3%	-0.63%
7	$16.00-$17.80	-1.2%	-0.56%
8	$18.00-$19.80	-1.0%	-0.49%
9	$20.00-$21.80	-0.9%	-0.45%
10	$22.00-$23.80	-0.8%	-0.41%
11+	over $24.00	< -0.7%	< -0.35%

Imagine a race with only two horses and two bettors. If each horse receives a one thousand dollar bet, its payoff would be the same: 1-to-1, assuming no takeout. But in reality, the track operator or bookmaker takes fifteen percent from the pool, so there's seventeen hundred dollars remaining to pay the winner. So the payoff odds would be 7-to-10 (a one thousand dollar bet would win seventeen hundred dollars). But there's also "breakage." The odds are rounded down to 3-to-5 (a one thousand dollar bet wins sixteen hundred dollars), and the track keeps the rounding error.

Say now that a third bettor places a three thousand dollar bet on one horse, so that four thousand dollars is bet on one horse and only one thousand dollars on the other. With no takeout, the respective payoff odds change to 1-to-4 and 4-to-1. But when the track removes its takeout that leaves four thousand, two hundred and fifty dollars. After breakage, the odds reduce to the minimum payoff of 1-to-20 and to 3-to-1.

As you can see, the takeout and breakage greatly reduce the expected gain of the bettors. The bets are taken, and the takeout, breakage, and odds are calculated and displayed, by the machine known as the "totalizator," invented by the Australian George Julius in 1912 as an entirely mechanical computer. Totalizators, popularly known as "tote boards," made their first appearance in the United States in the 1930s. They are now fully electronic computers capable of handling a variety of exotic betting combinations.

Independent off-track bookmakers ("bookies") have a smaller takeout and thus offer better payoffs, but they are illegal. However, with the introduction of Internet gambling have come "rebate shops," which are fully legal in their residence in such places as Bermuda. Whether it is legal for Americans to conduct transactions with the shops over the Internet is under debate. These off-shore betting shops return some percentage of every bet made to the bettor. By offering rebates, the shops are in effect reducing the takeout from twenty percent to as little as one or two percent. The shops can afford to reduce their takeout because they operate with minimal overhead. Low-takeout rebate shops allow skilled horseplayers who can afford to place expensive bets to make a steady income.

Winning with the Horses

Given the nature of the pari-mutuel system, are there any methods which allow opportunities for the skilled player? Most systems rely on using knowledge about the horses, jockeys, and race to pick the winners. These are not good systems, because the oddsmakers are experts with better access to the necessary information, and they shave the odds. And then there is the problem that the pari-mutuel system punishes making the most reasonable choices by reducing payoffs on favorites.

This is not to say that someone who has amazing horse sense can't do well, but nearly every person who's merely good (or very good) slowly (or not) and steadily loses money, unless he uses the rebate shops mentioned above. As with players in casinos, bettors are paying to watch horse racing—not a bad entertainment. However, there are more productive systems than trying to outguess the experts. One highly unethical and illegal but wildly successful method is to deliberately skew the pari-mutuel system in what is known as the builder play.

The builder play

The builder play is a scheme in which a group of conspirators monopolize the mutuel windows at the track to skew the payoffs by placing many small bets on longshots. Other conspirators place large off-track bets on the likely winners. The 1932 coup at the Agua Caliente track in Mexico increased the odds on winning horse Linden Tree from 7-to-10 to 10-to-1. The 1964 play at Dagenham Greyhound Stadium in East London controlled the bets such that only one ticket was sold on the winning quinella (Buckwheat and Handsome Lass) at odds of 9872-to-1.

Overlays

Your only legitimate chance to beat the pari-mutuel system is to find "overlays," that is, horses for which the track odds (determined by the masses at racetime) are better than the correct odds. If you know the real probabilities of horses winning, you can take advantage of the differences in the real odds and the track odds. Formulas exist to calculate maximal bets to exploit favorable odds without causing the pari-mutuel system to adjust against your edge. You can try to determine the correct odds through your own handicapping, but you're not likely to do better than the expert bookies who set the initial raceday odds.

Overlays exist because of quirks of mass psychology: the masses will bet disproportionately on long-odd horses (especially on the last race of the day), which draw down their odds, making them worse bets. Similarly, they tend to overbet on the favorite in early races. So the middle of the pack often end up with overlays. Note that these are not the long-odd horses, so there's no chance of a big payoff, and they're still less likely to win than the favorite. You have the classic gambling situation where your long-term expected gain is positive but you can expect to contend with plenty of short-term losses. Furthermore, if enough people take advantage of the favorable situation, the overlay quickly disappears.

An analytical method to identify the overlays based on mass psychology is to track what is the expected gain spread for various odds distributions in races. How the bets are distributed on different horses determines the distribution of payoff odds, the set of odds displayed on the tote board. Using a computer, it's possible to collect and interpret enough data on odds distributions to identify favorable betting situations. The central insight that distinguishes this number-crunching technique from others is that the behavior of the crowd of horse bettors is more predictable than that of the horses.

Sources

Aldous, D. and P. Diaconis. 1986. Shuffling cards and stopping times. American Mathematical Monthly 93 (May):333.

Bass, Thomas. *Eudaemonic Pie.* Houghton Mifflin, 1985.

Bass, Thomas. *The Predictors.* Henry Holt & Co., Inc., 1999.

Bayer, D. and Persi Diaconis, "Trailing the dovetail shuffle to its lair," Ann. Appl. Probab. 2(1992), no. 2, 294—313.

Berger, P. 1973. On the distribution of hand patterns in bridge: Man-dealt versus computer-dealt. Canadian Journal of Statistics 1:261.

Brunson, Doyle. *Doyle Brunson's Super System.* Cardoza Publishing, 1979.

Epstein, Richard A. *The Theory of Gambling and Statistical Logic, Revised Edition.* Academic Press, 1977.

Mann, B. "How many times should you shuffle a deck of cards?" UMAP J. 15 (1994), no. 4, 303—332.

Sklansky, David. *The Theory of Poker.* Two Plus Two Publishing, 1987.

Thorp, Edward. *Beat the Dealer.* Random House, Inc., 1962.

Thorp, Edward and Sheen Kassouf. *Beat the Market.* Random House, 1967.

Trefethen, L. N., and L. M. Trefethen. How many shuffles to randomize a deck of cards? Proceedings of the Royal Society, London A 456(Oct. 8, 2000):2561.